T0358321

CHARLES DARWIN

Charles Darwin's theory of evolution was one of the most significant revolutions in the history of science. Widely debated after the publication of *On the Origin of Species* in 1859, it continues to be controversial. In this volume, Michael Ruse offers the definitive history of the theory of evolution through natural selection. Tracing Darwin's intellectual journey and experiences that led him to his novel insights, Ruse explores his scientific contributions, their relationship to philosophical issues, and their religious implications, as well as how they have been both inspiration and challenge to novelists and poets. He also shows how Darwin's ideas continue to have contemporary relevance, as they shed light on social issues and problems such as race and sexual orientation – including the possibility and desirability of social change – the connections between Darwin's thinking and that of Sigmund Freud, and the status of women. Written in an engaging, non-technical style, Ruse's volume serves as an ideal introduction to the ideas of one of the key figures in the history of modern science.

MICHAEL RUSE recently retired from the position of Lucyle T. Werkmeister Professor of Philosophy at Florida State University and previously taught in the Department of Philosophy, University of Guelph, Canada. A scholar of Charles Darwin and the revolution associated with his name, he is the author and editor of over seventy books, founder of the journal *Biology and Philosophy*, a Guggenheim fellow, a Gifford lecturer, and a fellow of the Royal Society of Canada.

Figure 0.1 Charles Robert Darwin (1809–1882).

Charles Darwin

No Rebel, Great Revolutionary

MICHAEL RUSE
University of Guelph
Florida State University

Shaftesbury Road, Cambridge CB2 8EA, United Kingdom

One Liberty Plaza, 20th Floor, New York, NY 10006, USA

477 Williamstown Road, Port Melbourne, VIC 3207, Australia

314–321, 3rd Floor, Plot 3, Splendor Forum, Jasola District Centre,
New Delhi – 110025, India

103 Penang Road, #05–06/07, Visioncrest Commercial, Singapore 238467

Cambridge University Press is part of Cambridge University Press & Assessment,
a department of the University of Cambridge.

We share the University's mission to contribute to society through the pursuit of
education, learning and research at the highest international levels of excellence.

www.cambridge.org
Information on this title: www.cambridge.org/9781009438957

DOI: 10.1017/9781009438971

When citing this work, please include a reference to the DOI 10.1017/9781009438971

First published 2024

A catalogue record for this publication is available from the British Library

Library of Congress Cataloging-in-Publication Data
NAMES: Ruse, Michael, author.
TITLE: Charles Darwin : no rebel, great revolutionary / Michael Escott
Ruse, Florida State University
DESCRIPTION: Cambridge, United Kingdom ; New York : Cambridge University
Press, 2024. | Includes bibliographical references and index.
IDENTIFIERS: LCCN 2024006774 | ISBN 9781009438957 (hardback) |
ISBN 9781009438971 (ebook)
SUBJECTS: LCSH: Evolution (Biology) – History. | Evolution
(Biology) – Religious aspects. | Religion and science. | Darwin, Charles,
1809–1882 – Influence. | Biology – Philosophy.
CLASSIFICATION: LCC QH361 R86 2024 | DDC 570.9–dc23/eng/20240212
LC record available at https://lccn.loc.gov/2024006774

ISBN 978-1-009-43895-7 Hardback
ISBN 978-1-009-43894-0 Paperback

For My Graduate Students

Contents

Figures

Acknowledgments

As I wrote this book, on every page I felt the influence of my long-time mentor and friend, the late David Hull, Professor of Philosophy at Northwestern University. I could never thank him enough back then in the 1960s and 1970s, and I feel the same way today. Phillip Honenberger (2018) has written a remarkably sympathetic account of how David Hull and I worked separately on such issues as the nature of species, sometimes agreeing, more often disagreeing, but always in a very constructive manner. Anyone reading the piece will at once see that I would not be where I am today without the generous interactions with David Hull. I also owe much to the historian of nineteenth-century evolutionary thought Robert J. Richards, at the University of Chicago; and to the historian of twentieth-century evolutionary thought Joseph Cain, at University College London. Sydney Smith, biologist at Cambridge University, sharp as a whistle behind the façade of being the archetypical old English buffer, generously shared an immense knowledge of Darwin. Robert M. Young and Martin Rudwick were the two who, during my first sabbatical (at Cambridge) in the early 1970s, tutored me into being a real historian of science and not just a dilettante. I joke that I rarely agreed with anything Bob wrote or said and he agreed with nothing I wrote or said. It wasn't really a joke. The philosopher in me is deeply grateful for our interactions.

I am hugely indebted to Beatrice Rehl, my editor at Cambridge University Press. She is a worthy successor to the late Terry Moore at the Press, who so guided my early years. People such as I, who write and publish a great deal, are ever conscious that a good editor really should be acknowledged as a coauthor. I thank also Michael Reiss, Professor of Science Education at University College, London, who (at the time unknown to me) served as the final external reader of my manuscript. And last, but very much not least, the people who turned my somewhat shambolic manuscript into a book of the quality expected by Cambridge University Press: Senior Editorial Assistant Edgar Mendez; Nicola Maclean, Content Manager, Humanities; Project Management Executive, Veena Ramakrishnan; copyeditor Simon Fletcher; indexer Sergey Lobachev; my proofreader, Susan Olin; my illustrator, Martin Young.

As always, my wife Lizzie was there to encourage and support me. The same is true of my much-loved Cairn terriers, Scruffy McGruff and Duncan Donut. Wife and dogs are good at making me close my laptop and go for a walk. Finally, the dedication is to all my former graduate students, in Canada and in the United States. As you read this book, you will realize it is written as much by a teacher as by a scholar. It is for this reason I appreciate especially the choice of Michael Reiss as my outside reader. I am indeed a fortunate man to have had such an integrated life, doing the two things that mean so much to me.

Introduction

> The nature of physical things is much more easily conceived when they are beheld coming gradually into existence, than when they are only considered as produced at once in a finished and perfect state.
>
> John Dewey, quoting René Descartes, *Discourse on Method*

More than four decades ago, I wrote *The Darwinian Revolution: Science Red in Tooth and Claw*. For all that the eminent evolutionist Ernst Mayr chided me for the silliness of my subtitle, I remain very proud of that book. As might be expected, much of what I wrote then is seriously dated, I would like to think in major respects because of the work that book stimulated – work by myself and others. It has long been my hope that, as my career of over fifty years as a philosopher and historian of science draws to an end, I could write a serious revision of the book that helped launch my career.

This is that revision. Except it isn't really. Most importantly, the very intent of the earlier book has been changed, and this (not Mayr) is the reason for the change of title. Then, I wrote a straight history of science, trying to show what happened in the Darwinian Revolution. It was a much needed overview, much needed because of the flood of new information and ideas that had appeared in the twenty years since the history of science became professionalized. It was the book I wished I had had ten years before, when, as a young philosopher of science under the influence of Thomas Kuhn, I turned to the history of science. As one who, in childhood, prayed for wet weather so the order to go out and play was rescinded and I could finish reading *The Children of the New Forest* and move on to *The Secret Garden*, Darwin and his achievements were from a time and a land where it was always raining – leading me in a direction I have never regretted.

Now my commitment to philosophy has reasserted itself, and this book here is a history of ideas. By this I mean, writing in the tradition of Arthur Lovejoy and Isaiah Berlin, I am using history to throw philosophical light on issues that engage us today. The book is deeply autobiographical. It is by no means simply a précis of work I have done in the past forty years. However, unashamedly, I will use already expressed ideas to push forward to my concerns now. Specifically, I shall ask about the relevance of Darwin's work towards an understanding of attitudes towards foreigners, especially immigrants; towards an understanding of the nature (if they exist) of racial differences, and how these (real or otherwise) affect society's attitudes towards African-Americans; towards an understanding of sexual orientation, whether it is a matter of nature or of choice; and, finally, towards an understanding of the nature and status of women. Recently, it has become evident that there is still huge prejudice against Jews. After I have discussed beliefs about foreigners and attitudes towards race, I add a short codicil addressing this issue. Overall, I shall look at Darwin's work against its background, at our thinking today and the extent it has been shaped by Darwin's work, and whether Darwin himself had any idea of the ways in which his findings and theories would be an integral part of our thinking today. The proof of the pudding is in the eating. Here, I will not defend my change of intent. The reader must judge whether the change was proper and whether I have succeeded in what I have set out to do.

I will say, however, that I write within a framework – more precisely, against a framework. In my earlier book, I acknowledged that, whatever the importance of Darwin's science, particularly in the *Origin of Species* and the *Descent of Man*, in respects he did not do what he set out to do, namely convince professional evolutionists of his own generation to adopt, as the chief mechanism of change, Darwin's cause: natural selection. I did not then see that this was a claim with supposed wider implications, namely that it is a mistake to think that Darwin led to an actual scientific "revolution." That he was rather one of many who contributed to the nineteenth-century change from a world of the miraculous origins of organisms to a work of the natural origins of organisms. In other words, while there was certainly a general nonevolutionary consensus before the *Origin*, and there was a general evolutionary consensus after the *Origin*, really Darwin had little or no role to play in the change. As they accept the literal resurrection of Jesus, the general public might accept the revolutionary nature of Darwin's legacy. Those in the know realize that neither claim withstands the critical eye. In the Darwinian case, given especially that Darwin's theory was already existing beliefs stitched together – in this respect he was certainly no rebel – talk of "revolutions" is pushing beyond the boundaries.

Typical of criticisms of the "revolutionary" claims for Darwin's achievement are the concluding words of James Secord at the end of his (deservedly) prize-winning book on the pre-*Origin* evolutionary work *Vestiges of the Natural History of Creation* by the Scottish publisher Robert Chambers. Darwin is important, but not that important. Many of the claims promoting his importance are "implausible." Adding: "the Origin's main novelty, natural selection, was rejected by almost all readers in the first seventy-five years after publication" (Secord 2000, 516). Secord is but one of a number of voices that want to shrink the author of the *Origin of Species* down to size. He and the others are nothing to Peter Bowler, the eminent historian of evolutionary biology. The titles of three of his books tell the tale: *The Eclipse of Darwinism* (1983); *The Non-Darwinian Revolution: Reinterpreting a Historical Myth* (1988); and *Darwin Deleted: Imagining a World without Darwin* (2013). That tells it like it is! Bluntly: "There is now a substantial body of literature to convince anyone that the part of Darwin's theory now recognized as important by biologists had comparatively little impact on late nineteenth century thought" (1988, ix).

"Comparatively little impact on late nineteenth century thought"?! Although, primarily, I am telling the tale of Darwin and his accomplishments, I write against the *background* of this claim and I look at the evidence that leads to such a judgment. Since the *Origin* is – or claims to be – a work of science, let us be generous and assume that it is to this that people such as Bowler would have us turn. So let us pick up the challenge. However, not to make hasty judgments, constrained by the interests of Bowler and other Darwin belittlers, I shall also look at other areas of inquiry that might have felt the effects of the arrival of the idea of natural selection – philosophy, religion, literature. Also, since the titles and contents of Bowler's books certainly suggest that he is talking of the Darwin Revolution without temporal restrictions, I shall reject the assumption that one can make a clean division between "revolutionary" in the nineteenth century and "revolutionary" in the twentieth century. These topics and interests one might regard as the foreground of my discussion.

Let us turn at once to see if I have succeeded in what I set out to do.

1

Beginnings

Organicism

Plato's writings were cast in the dialogue form, usually with the philosopher Socrates as the main figure, talking, teaching, arguing with his disciples. Over the years, Plato increasingly used this dialogue form to introduce his own ideas, putting them in the mouth of Socrates. One such dialogue, the *Phaedo*, purports to tell of the last day of Socrates, before he is forced to drink poison, a punishment for filling his young admirers with all sorts of treasonable ideas. Plato has Socrates tackle the question of the possible chance nature of the universe, a problem of pressing importance to one about to die, having Socrates argue that truly all must be the product of a designing intelligence. "One day I heard someone reading, as he said, from a book of Anaxagoras, and saying that it is Mind that directs and is the cause of everything. I was delighted with this cause and it seemed to me to be good, in a way, that Mind should be the cause of all" (Cooper 1997: *Phaedo* 97, c–d).

In another dialogue, the *Republic*, Plato fit this idea into his overall metaphysical picture of reality. The main aim of this dialogue is to set up the ideal society, one that he thinks is based on our realization that this makes for the happiest form of life. The rulers – the "philosopher kings" – will be guided by their understanding of the nature of reality. This world of ours is the world of change, of becoming. It is not unreal, but it only reflects the world of ultimate reality, the unchanging world of the Forms. These are universals, standards, that guide and inform our world of experience. Dobbin is an individual horse. Dobbin is a horse, not a dog, because he "participates" in the Form of Horse. Fido, the family dog, participates in the Form of Dog. These forms are hierarchical, linked together through their relationship to the ultimate form, the Form of the Good. It is this that is in some sense the guiding intelligence. The equivalent in our world is the sun, which likewise has the role of linking all together and making possible continuation and thriving. First it illuminates:

> Light is the noble bond between the perceiving faculty and the thing perceived, and the god who gives us light is the sun, who is the eye of the day, but is not to be confounded with the eye of man. This eye of the day or sun is what I call the child of The Good, standing in the same relation to the visible world as The Good to the intellectual. (Cooper 1997, 508c–509a)

And then it is the sustenance, as one might say, that leads to growth: "And this Idea of Good, like the sun, is also the cause of growth, and the author not of knowledge only, but of being, yet greater far than either in dignity and power."

It is in a later dialogue, the *Timaeus*, that Plato argued for an organismic view of the universe – the organism was the *root* metaphor – with The Good being characterized as the "Demiurge." This Creator made the world an organism, so that it could be as good, as perfect, as possible. It is valuable:

> God desired that all things should be good and nothing bad, so far as this was attainable .... For which reason, when he was framing the universe, he put intelligence in soul, and soul in body, that he might be the creator of a work which was by nature fairest and best. Wherefore, using the language of probability, we may say that the world became a living creature truly endowed with soul and intelligence by the providence of God. (Cooper 1997)

Aristotle, Plato's student, was also an organicist, with a very different take from that of Plato. For a start, unlike Plato, he did not think that universals were entities existing in their own right, in a transcendent world of Forms. He thought rather that universals were more like templates, and they had existence only in the individuals of this world. Dobbin and Daisy were formed in the same pattern, and there is nothing beyond this. Again, Aristotle did not believe in an external Designer. He believed in something Godlike – the Perfect Being. This is not a physical being, but in some sense thought personified. "For that which is capable of receiving the object of thought, i.e. the essence, is thought. But it is active when it possesses this object." Hence, life "belongs to God; for the actuality of thought is life, and God is that actuality; and God's self-dependent actuality is life most good and eternal. We say therefore that God is a living being, eternal, most good, so that life and duration continuous and eternal belong to God; for this is God" (Barnes 1984: *Metaphysics*, 12, 1072b).

Famously, Aristotle divided causes into four categories (*Physics*, 194b16–195a3). Suppose we want to make a statue, for example of a British private – a "Tommy" – from the First World War (Reiss and Ruse 2023, 17) (Figure 1.1). You start with the *efficient* cause, the modeler or sculptor who actually made the statue. Then next you have the *material* cause, the substance from which

Figure 1.1 Statue of a British WWI soldier, a "Tommy."

it is made – metal (bronze) or stone (marble) or whatever substance. Then you have the *formal* cause, the pattern that Plato was trying to capture with his theory of forms. The model must look like a real British soldier. It would not be wearing a hat with a *Pickelhaube* for instance. And then, fourth, in a way the most important of all, you have the *final* cause. The teleological element behind your commissioning the statue. Why is it being made? The answer is simple. Future generations will be alerted to, and give thanks for, the sacrifices of such humble men and their comrades.

One of the problems with teleology, final-cause thinking, is that of the "missing goal object." If you hear someone hammering away, you can easily identify the efficient cause. It is a hammer striking a nail as it penetrates a plank being laid down as a floor. Material causes are iron and wood. Formal cause is the kind of house you are intending to build – a row house, semi-detached house, bungalow, or whatever. Final cause is the yet-to-be erected house. The final cause of the statue of the soldier, in the middle of the village green, is (as just noted) to remind us each time we pass by of the sacrifices made by so many young men in the Great War so that we might live in harmony and peace. But what if, halfway through your building, you fail to get planning permission and you have to tear everything down? What if there is an accident when transporting the statue to the village, it is destroyed, and the parish simply does not have the money to replace it? They are going to have to be satisfied with a brass plaque. How can we speak of final cause when it never happens? Plato has a ready answer. The final cause is the thought of the house, of the statue. It is in fact a kind of mental efficient cause. There is no such easy way out for Aristotle. He has to say something like, there is a force, a tendency, directed toward the house or the statue. This exists now so is a kind of efficient cause, and it simply doesn't get to its end. The direction exists now.

Final causes must be saved. Final causes can be saved. So, we can still ask, meaningfully: Where do humans come in all of this? As you might expect, at the top! We are the animal equivalent of the mighty oak. Monad to man. We are the ultimate final cause. There is direction, from lesser to greater, from (and this is important) little worth or value to greater worth or value. Note that this is worth or value that is objectively "out there." It is not a judgment based solely on our preferences or desires. I am a passionate supporter of the Wolverhampton Wanderers soccer club, "the Wolves." Regretfully, these days this is rarely something based on objective value. To the contrary, Plato tells us: "God gave the sovereign part of the human soul to be the divinity of each one, being that part which, as we say, dwells at the top of the body, inasmuch as we are a plant not of an earthly but of a heavenly growth, raises us from earth to our kindred who are in heaven" (Cooper 1997, 90b). Likewise, Aristotle: "after the birth of animals, plants exist for their sake, and that the other animals exist for the sake of man … . Now if nature makes nothing incomplete, and nothing in vain, the inference must be that she has made all animals for the sake of man" (Barnes 1984, 1256b15–22). The unique bipedality of humans is also readily understood: "of all living beings with which we are acquainted man alone partakes of the divine, or at any rate partakes of it in a fuller measure than the rest." Hence, "in him alone do the natural parts hold the natural position; his upper part being turned towards that which is upper in the universe. For, of all animals, man alone stands erect" (656a17–13).

Although, as standing outside the Judeo-Christian tradition, technically Plato and Aristotle qualify as "pagans," the last thing that would have appealed to either would have been dancing stark naked save for Birkenstocks, around a campfire, out in California. (Socrates might have welcomed the chance, so long as his fellow dancers were attractive young men.) To the contrary, the seminal Christian thinkers – Augustine and Aquinas, particularly – were greatly influenced by the Greeks. This, despite the fact that neither read

Greek. Augustine got his understanding from the Neoplatonist Plotinus. Aquinas reaped the rewards of recent translations (into Latin) of original Greek texts, particularly those of Aristotle.

In his *Confessions*, Augustine's characterization of God could have come straight out of the *Republic*. Necessary: "For God's will is not a creature but is prior to the created order, since nothing would be created unless the Creator's will preceded it. Therefore, God's will belongs to his very substance." Outside space: "no physical entity existed before heaven and earth." Outside time: "Your 'years' neither come nor go. Our years come and go so that all may come in succession. All your 'years' exist in simultaneity, because they do not change; those going away are not thrust out by those coming in … Your Today is eternity." Likewise, the design and creation of the Earth.

> Even leaving aside the voices of the prophets, the world itself, by the perfect order of its changes and motions, by the great beauty of all things visible, claims by a kind of silent testimony of its own both that it has been created, and also that it could not have been made other than by a God ineffable and invisible in greatness, and ineffable and invisible in beauty. (Augustine 396, *Confessions*, 53)

Ours is a world of great value, coming from God. "And God saw everything that he had made, and, behold, it was very good" (Genesis 1:31). Humans, one hardly need say, are "very, very good." "Thou sayest not, 'Let man be made,' but Let us make man. Nor saidst Thou, 'according to his kind'; but, after our image and likeness" (Augustine 396, *Confessions*, 13).

As one influenced by Aristotle, Aquinas tended more to an internal reading of final cause, but the message was the same. Their very functioning shows that living things are of great value and humans of the greatest value. For a Christian, faith will always outrank reason. Remember the story of Thomas, who was scolded for demanding evidence that the man before him was indeed the crucified Christ. But reason is crucially important. Nicely backing Aquinas's conviction that reason does point to God is the fact that Aristotle embraced a geocentric view of the universe (with Earth at the center). This was very much in line with what Aquinas wanted to believe. The Earth is not just another planet, but (literally) the center of the universe, where all the action takes place. "The heavens are moved by God, and they in turn affect what happens down here on Earth" (Aquinas 1947, *Compendium Theologiae* I, 4).

> All motion is observed to proceed from something immobile, that is, from something that is not moved according to the particular species of motion in question. Thus we see that alterations and generations and corruptions occurring in lower bodies are reduced, as to their first mover, to a heavenly

body that is not moved according to this species of motion, since it is incapable of being generated, and is incorruptible and unalterable.

Mechanism

Back in the time of the Ancient Greeks, there were those who were unimpressed by the organic metaphor. They saw the world as meaningless, in the sense that there was no organizing force, internal or external. No values. Everything was the result of one thing happening after another. One set of particles, "atoms," existing in otherwise empty space, the "void," reconfiguring themselves driven by blind law. Given enough time, given enough combinations, and things would begin to work. Even before Socrates, the atomists – Leucippus, Democritus, and a little later Epicurus – were denying final cause and putting everything down to efficient cause. This got its fullest expression in the work of the pre-Christian Roman poet Lucretius (1950). Laying things out in his *De Rerum Natura* (*On the Nature of Things*), he made the case that all was a product of chance, with no direction.

> At that time the earth tried to create many monsters
> with weird appearance and anatomy –
> androgynous, of neither one sex nor the other
> but somewhere in between; some footless, or handless;
> many even without mouths, or without eyes and blind;
> some with their limbs stuck together all along their body,
> and thus disabled from doing harm or obtaining anything they needed.
> These and other monsters the earth created.
> But to no avail, since nature prohibited their development.
> They were unable to reach the goal of their maturity,
> to find sustenance or to copulate.
>
> (Sedley 2007, 150–53, *De rerum natura* V 837–848)

Then, from grotesque figures – three legs, one coming in the middle of the back, no mouth or eyes but several pairs of ears, and more – slowly functioning creatures started to appear.

> First, the fierce and savage lion species
> has been protected by its courage,
> foxes by cunning, deer by speed of flight.
> But as for the light-sleeping minds of dogs, with their faithful heart,
> and every kind born of the seed of beasts of burden,
> and along with them the wool-bearing flocks and the horned tribes,
> they have all been entrusted to the care of the human race, ...
>
> (V 862–867)

No final causes, only efficient causes. Eyes just appeared, and then they were put to use. To think otherwise is to get things backwards.

> All other explanations of this type which they offer
> are back to front, due to distorted reasoning.
> For nothing has been engendered in our body
> in order that we might be able to use it.
> It is the fact of its being engendered that creates its use.
>
> (V 832–835)

Expectedly, especially given the coming of Christianity, none of this convinced. It was at most a curiosity – an example of how not to use one's reason. No matter how many typewriters, monkeys do not produce Shakespeare.

Then, around 1500, things started to change: With the Renaissance came a whole new appreciation of the thinking of the past, especially pre-Christian thinking. Writings such as *On the Nature of Things* were hauled out and studied in their own right. Paralleling the Renaissance was the Reformation, when Martin Luther, followed by Jean Calvin and Huldrych Zwingli, broke from Rome and started the Protestant challenge. There are many ways of categorizing this major break, but above all it was a move from the overintellectualized Catholic form of Christianity – epitomized by the theology of Aquinas – to a more literal form of religion. A religion, based on the Bible – *sola scriptura* – undergirded by faith rather than reason. Famously, or perhaps notoriously, Luther said: "Reason is a whore, the greatest enemy that faith has; it never comes to the aid of spiritual things, but more frequently than not struggles against the divine Word, treating with contempt all that emanates from God" (Luther 1914, 51, 126, 7). This was not a critique of organicism as such, but it was a philosophy that did not regard organicism as God's way of thinking, as one might put it.

Third and most important of all was the Scientific Revolution, from the heliocentric universe of Copernicus's De revolutionibus orbium coelestium (*On the Revolutions of the Celestial Spheres*) (1543), to Newton's theory of gravity, Philosophiæ Naturalis Principia Mathematica (*Mathematical Principles of Natural Philosophy*) (1687). More than just raw science, it was a change of *root* metaphors, from the organism to the *machine*.

> At all times there used to be a strong tendency among physicists, particularly in England, to form as concrete a picture as possible of the physical reality behind the phenomena, the not directly perceptible cause of that which can be perceived by the senses; they were always looking for hidden mechanisms, and in so doing supposed, without being concerned about this assumption, that these would be essentially the same kind as the simple instruments which men had used from time immemorial to relieve their

> work, so that a skillful mechanical engineer would be able to imitate the real course of the events taking place in the microcosm in a mechanical model on a larger scale. (Dijksterhuis 1961, 497)

But why did this metaphor conquer? In major part because people were starting to invent and use machines. The clock above all: "it is no less natural for a clock constructed with this or that set of wheels to tell the time than it is for a tree which grew from this or that seed to produce the appropriate fruit." Reduction! Take it to bits and see how it works. "Men who are experienced in dealing with machinery can take a particular machine whose function they know and, by looking at some of its parts, easily form a conjecture about the design of the other parts, which they cannot see." Same for organisms. "In the same way I have attempted to consider the observable effects and parts of natural bodies and track down the imperceptible causes and particles which produce them" (Descartes, *Discourse on Method* [1637] 1985, 288–89). Note that a natural consequence of this way of thinking is that the notion of "mechanism" has two connected but distinct meanings. One refers to the root metaphor – the world overall is to be considered in machine terms. It is like one big functioning entity. The other refers to individual cases and causes. The mechanism driving a watch consists of springs and so forth connected in particular ways to make the hands circulate in a steady, regular manner. Generally, there is no ambiguity about the particular sense in which the term is being used.

Where Descartes led, others followed. In his *A Free Enquiry into the Vulgarly Received Notion of Nature*, Robert Boyle, seventeenth-century chemist and philosopher, set things out (Figure 1.2). The world is

> like a rare clock, such as may be that at Strasbourg, where all things are so skillfully contrived that the engine being once set a-moving, all things proceed according to the artificer's first design, and the motions of the little statues that at such hours perform these or those motions do not require (like those of puppets) the peculiar interposing of the artificer or any intelligent agent employed by him, but perform their functions on particular occasions by virtue of the general and primitive contrivance of the whole engine. (Boyle 1996, 12–13)

Most immediately, the heliocentric universe envisioned by Copernicus simply has objects going round and round according to unbroken law. Just like a clock. One might think there is an ultimate purpose. A clock tells the time. The universe provides a home for humankind. But this is not part of the scientific explanation. Unlike the organic metaphor, there are no objective values presupposed by this overarching metaphor. No values. Where then does God figure in all of this? He was a "retired engineer" (Dijksterhuis 1961, 491). In a

Figure 1.2 The clock in Strasbourg Cathedral, 1574. (This is the updated version of 1843.)

way this is true. We are moving from the organicist's God, always at work, to a more distant God, who sets things going and then stands back.

> The ultimate mystery resided in God rather than in Nature, which could thus, by successive steps, be seen not as a self-sufficient Whole but as a divinely organized machine in which was transacted the unique drama of the Fall and Redemption. If an omnipresent God was all spirit, it was all the more easy to think of the physical universe as all matter; the intelligences, spirits, and Forms of Aristotle were first debased and then abandoned as unnecessary in a universe that contained nothing but God, human souls, and matter. (Hall 1954, xvi–xvii)

Note that this does not mean that God is nonexistent, although people certainly thought Lucretius veered that way. But it does push one toward what is known as "deism" – God created and then stood back – as opposed to "theism" – the God of the Christian always involved in His creation. In the words of the eighteenth-century thinker Matthew Tindal, Jesus was simply telling us that everything is ruled by unbroken law. "The Religion of the Gospel is the true original Religion of Reason and Nature" (1730, 6).

Evolution

As we move on past the Scientific Revolution, we enter the Age of the Enlightenment. Until then, Christianity had been a religion where we are always dependent on God. Thanks to Adam and Eve, we are all tainted by sin, and, except through the sacrifice on the cross, we are lost. This is made clear in the much loved hymn of Isaac Watts (1707).

> When I survey the wondrous cross
> On which the Prince of Glory died
> My richest gain I count but loss
> And pour contempt on all my pride

Providence!

Now, in the opinion of Enlightenment thinkers, we must rely on ourselves. God is not going to intervene. But, thanks to our discoveries – in medicine, in engineering, in running our lives – we can improve our lot.

Progress!

And very soon people began to wonder if progress was more widespread. God is not going to create things miraculously. This must be left to the laws of nature. But can we slip in values even in a mechanist's view of the world? The French *philosophe* Denis Diderot is perhaps best known for his pornographic writing about lesbian nuns.

> The hand she had rested on my knee wandered all over my clothing from my feet to my girdle, pressing here and there, and she gasped as she urged me in a strange, low voice to redouble my caresses, which I did. Eventually a moment came, whether of pleasure or of pain I cannot say, when she went as pale as death, closed her eyes, and her whole body tautened violently, her lips were first pressed together and moistened with a sort of foam, then they parted and she seemed to expire with a deep sigh. (Diderot [1796] 1972, 137–38)

Expectedly, one who had so freed himself from the constraints of conventional religion felt open to speculate about the possibility of progress working its way in the world of animals and plants. "Just as in the animal and vegetable kingdoms, an individual begins, so to speak, grows, subsists, decays and passes away, could it not be the same with the whole species?" Continuing:

> would not the philosopher, left free to speculate, suspect that animality had from all eternity its particular elements scattered in and mingled

> with the mass of matter; that it has happened to these elements to reunite, because it was possible for this to be done; that the embryo formed from these elements had passed through an infinity of different organizations and developments.

Indeed, who dare now say that everything is over? Life

> has perhaps still other developments to undergo, and other increases to be taken on, which are unknown to us; that it has had or will have a stationary condition; ... that it will disappear for ever from nature, or rather it will continue to exist in it, but in a form, and with faculties, quite different from those observed in it at this moment of time. (Diderot 1943, 48)

Playing the same song, on the other side of the Channel, was the English physician and part-time poet Erasmus Darwin (paternal grandfather of Charles). He too was given to somewhat risqué thoughts on sexuality, although he confined his attentions to plants rather than nuns.

> Ten brother-youths with light umbrella's shade,
> Or fan with busy hands the panting maid;
> Loose wave her locks, disclosing, as they break,
> The rising bosom and averted cheek;
>
> (Darwin 1789, Canto IV)

Turning to the nature of species, he gave full throat to the ideas that so excited Diderot.

> Organic Life beneath the shoreless waves
> Was born and nurs'd in Ocean's pearly caves;
> First forms minute, unseen by spheric glass,
> Move on the mud, or pierce the watery mass;
> These, as successive generations bloom,
> New powers acquire, and larger limbs assume;
> Whence countless groups of vegetation spring,
> And breathing realms of fin, and feet, and wing.
>
> Imperious man, who rules the bestial crowd,
> Of language, reason, and reflection proud,
> With brow erect who scorns this earthy sod,
> And styles himself the image of his God;
> Arose from rudiments of form and sense,
> An embryon point, or microscopic ens!
>
> (Darwin 1803, 1, 11, 295–314)

And lest there be any mistake, explicitly Darwin tied this in with thoughts of cultural progress. The idea of evolution "is analogous to the improving

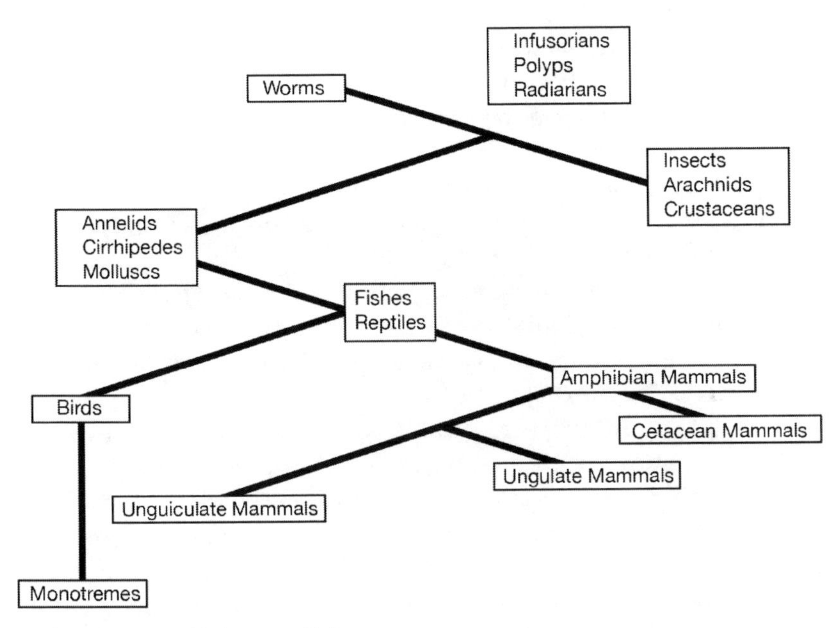

Figure 1.3 Lamarck's history of life.

excellence observable in every part of the creation; such as the progressive increase of the wisdom and happiness of its inhabitants" (Darwin [1794–96] 1801, 2:247–48). Value!

Best known of all, the French biologist, Jean Baptiste de Lamarck, writing at about the same time as Darwin – his masterwork, *Philosophie Zoologique*, was published in 1809 – seems to have presupposed some kind of Aristotelian vital force, teleologically directed toward humans (Figure 1.3). Note that Lamarck was not thinking in tree-metaphor terms. The higher animals are at the bottom rather than the top. He believed that life is continuously spontaneously generated, and all newly generated organisms follow the same path. Hence, if lions go extinct, we just have to wait and then more will appear, having followed the same path of those now gone. Famously, Lamarck is better known today for his secondary mechanism, one which now bears his name, the inheritance of acquired characteristics. The blacksmith's arm gets more muscular thanks to days at the forge, and so his children are born with muscular arms built-in, as it were. But even if this be true – another point we shall pick up on – it hardly explains full-blown evolution. Muscular arms do not explain the evolution of reptiles from fish, or mammals from reptiles.

Lamarck's neo-Aristotelianism is, in a way, more consistent than Erasmus Darwin's views. Darwin wants to slip in progress, a value notion, and hold on to a mechanist view of the physical world. Lamarck is more openly rejecting pure mechanism as a root metaphor. Yet this gives us the clue to the big difficulty. How, if we do not appeal to some kind of teleological vital force, do we explain the final-cause nature of organisms – fins, wings, legs, brains? We are cast right back to the Lucretius problem. Blind law simply does not make for design-like features. And if we opt for a Platonic rather than an Aristotelian solution, as a (root metaphor) mechanist one is no better off. True, God designed it all in the beginning, but we are still stuck with the blind-law problem. Robert Boyle spotted this. He embraced mechanism fully for the non-organic world. Planets circling the sun, qua science, have no purpose. The world is (objectively) value-free.

But there is still the problem of organisms. They seem to be value impregnated. Whether we care or not, the heart is of value to the mammal. And progress likewise brings in value, again whether we care or not. Boyle's solution was to kick this problem out of science and to say it was all a problem for religion. God did intervene directly when it comes to the final-cause nature of organisms. In other words, talk of mechanisms is part of science. Talk of final causes is part of theology! In his *Disquisition about the Final Causes of Natural Things*, Boyle wrote:

> For there are some things in nature so curiously contrived, and so exquisitely fitted for certain operations and uses, that it seems little less than blindness in him, that acknowledges, with the Cartesians [followers of the French philosopher Descartes], a most wise Author of things, not to conclude, that, though they may have been designed for other (and perhaps higher) uses, yet they were designed for this use. (Boyle 1688, 5: 397–98)

Continuing, that supposing that "a man's eyes were made by chance, argues, that they need have no relation to a designing agent; and the use, that a man makes of them, may be either casual too, or at least may be an effect of his knowledge, not of nature's." However, intermingling science and religion, the penalty from taking us away from a designing intelligence is taking us from the chance to do science – the urge to dissect and to understand how the eye "is as exquisitely fitted to be an organ of sight, as the best artificer in the world could have framed a little engine, purposely and mainly designed for the use of seeing" (5: 398).

So, Boyle tells us, we go from science to theology and onto God. First:

> In the bodies of animals it is oftentimes allowable for a naturalist, from the manifest and apposite uses of the parts, to collect some of the particular, to which nature destinated them. And in some cases we may, from the known natures, as well as from the structure, of the parts, ground probable conjectures (both affirmative and negative) about the particular offices of the parts. (5: 424)

Then, second, the science finished, theology steps up to the plate: "It is rational, from the manifest fitness of some things to cosmical or animal ends or uses, to infer, that they were framed or ordained in reference there unto by an intelligent and designing agent" (5: 428). In short, from a scientific study of what Boyle called "contrivance," in the domain of science, we go on to inferences about design – or rather Design – in the domain of theology. Thanks particularly to the influence of Aquinas, organisms were understood in an Aristotelian way. With the coming of mechanism as a root metaphor, Aristotelian vital forces were banished from science, and purpose, inasmuch as it applies to organisms, was seen as a consequence of Plato's Demiurge, now (thanks particularly to Augustine) identified as the Christian God.

A move from Aristotle to Plato. A solution. To us, it may not seem an altogether satisfactory solution. It is, however, important to place this in historical context. The English – not the Scots, who, thanks to Calvin's disciple John Knox, became hard-line Presbyterians – had a particularly idiosyncratic form of Protestantism. Their religion was a function, not of theological disputes, but of the fact that Henry VIII wanted to free himself from his marriage to Catherine of Aragon so he could marry Anne Boleyn. The pope wouldn't let him. So he picked up his country and went home. When Elizabeth, the daughter of Anne, became queen, she inherited a hybrid religion, with the fabric and ceremonies of the previous Catholicism melded with a mild Calvinistic theology. To cement their religion, for a country starting to rely more and more on machinery, emphasizing natural theology – reason and evidence – was a slam dunk. So, for Boyle and for those who followed him, like John Ray (the first in a long line of parson–naturalists and a pioneer in modern taxonomy), making much of God's designing influence was part of the package of being English. In a way, it was no compromise, and it did mean that on and through the eighteenth and into the nineteenth century, the English could go on doing what we today would call "biology" (Ruse 2003). Albeit, at the cost (as others would regard it) of acknowledging that, unlike the study of planets, we are no longer doing science.

Expectedly, we find continental philosophers worrying about this. The greatest of them all, Immanuel Kant, devoted a considerable section of his

Third Critique, The Critique of Teleological Judgement (1790), to this very problem. As an avowed Newtonian, Kant started with the premise that organisms are just machines. As one raised a Pietist (a kind of ultra-Lutheran), Kant wanted no truck with natural theology and the like. Given that science is the domain of reason and evidence, we cannot therefore put final cause down to God (even if, on faith, we may believe this). Kant's trick – "evasive strategy," if you like – was to regard final-cause thinking as a *heuristic* guide. Final causes help us think about organisms. They are "regulative." They are not part of reality.

One much influenced by Kant's ideas – to whose critical philosophy he "was greatly indebted" – was the French father of comparative anatomy, Georges Cuvier (Coleman 1964, 16). He was born in one of the border provinces that was not then incorporated into France – which accounts both for his having been German-educated and Protestant. Based on this Protestantism and reaffirmed by careful study of Aristotle, Cuvier made the Kantian insistence on teleology central to his approach to life, both living and dead. He stressed that, in considering an organism, we have to look at how the various parts fit and work together. We have to dig into the organization of the organism and ask about purposes. Justifying this, as it were, was something Cuvier called the "conditions of existence." This demands that we look at the parts of organisms from a final-cause perspective. "As nothing can exist without the reunion of those conditions which render its existence possible, the component parts of each being must be so arranged as to render possible the whole being, not only with regard to itself but to its surrounding relations." What does this imply for those who would try to understand things? "The analysis of these conditions frequently conducts us to general laws, as certain as those that are derived from calculation or experiment" (Cuvier 1817, 1, 3–4).

We must keep value questions in front of us all the time. What is the purpose of a particular part? It is hard to overestimate the importance of this principle for nineteenth-century biology. The British particularly seized on it as something exactly paralleling their natural-theological concerns. For Cuvier, for all that he was a Protestant, he was first a Frenchman in the tradition of Descartes. He wanted no part of the Creator in his science. For the British, however, it was important to show that their work was very much in line with proper (English) theological thinking (Ruse 1979). They still stood in the tradition going back to Boyle and earlier, and – especially since many of the scientists were fellows of Oxbridge colleges and hence necessarily had to be ordained members of the established (Anglican) church – they needed a shield against those who claimed that empirical inquiry can lead only to infidelity and heresy, and worse. When the greatest biologist of the time thrust final-cause thinking forward, it was manna from heaven, to coin a phrase. Now, someone like William Whewell (pronounced "Hule"), historian and philosopher of science and future master of Trinity College, Cambridge, had the perfect escape from heretical visions like that of Erasmus Darwin or Lamarck. The problem of final cause meant that, seeing science as Newtonian and hence under the machine root metaphor, there could be no scientific solution to the problem of organic origins. "Science says nothing, but she points upwards" (Whewell 1837, 3, 588).

"Manna from heaven." Manna, as we shall see in Chapter 2, that when fermented, somewhat unexpectedly and not entirely happily, proved to have the kick of a horse.

2

Charles Robert Darwin

Darwin before Evolution

Charles Robert Darwin was born on February 12, 1809, the same day as Abraham Lincoln across the Atlantic (Browne 1995, 2002). He died on April 19, 1882. He was the fifth of six children, three girls (Marianne 1798–1858; Caroline 1800–88; Susan 1803–66), then two boys, Charles and his older brother Erasmus (1804–81), and finally a sixth child, the fourth daughter (Catherine 1810–66). His father, Robert Darwin (1766–1848), one of the many children (at least fourteen) of the evolutionist, was a physician in the British Midlands – in Shrewsbury, on the border of Wales. His mother (Susannah 1765–1817) was the daughter of Joseph Wedgwood, the founder of the pottery manufacturing company and friend of the older Erasmus Darwin. She died when Charles – known to the family as "Bobby" – was eight years old. He was raised, with great emotional warmth, by his older sisters. The Darwin family were Anglicans (Church of England), although it was clear that Robert's nonbelief would have given Richard Dawkins pause. It is thought that the sisters were more Low Church in their inclinations. Charles Darwin's future wife, his first cousin Emma Wedgwood (1808–96), was raised in the Church of England, although the Wedgwood family was empathetic to the Unitarians – meaning they denied the Trinity and regarded Jesus as a very good man but not divine. (Darwin's mother was raised a Unitarian.) Both families were united in a hatred of slavery (Figure 2.1). This was not a universal opinion in Britain, especially in the North, where the factories depended on cotton grown in the southern slave states of the Union. Robert Darwin was a very successful physician, although the main source of his (very large) income came from raising loans. As a medical man, who met the widest range of people, he was perfectly positioned to persuade financiers to put up money for mortgages on the estates of impecunious landowners. More money – much

Figure 2.1 Wedgwood anti-slavery medallion, 1787.

more money – came from the Wedgwoods, whose works was one of the most successful in then-industrializing Britain. To give an indication of the family wealth, in 1881, a year before he died, as a New Year's gift, Darwin gave each of his children (he had ten, seven of whom lived to maturity) £500. The novelist Anthony Trollope estimated that he could raise a (middle-class) family on £400 a year, another £100 if he wanted to go fox hunting (he did).

From an early age, both Darwin boys realized they would not have to work for a living. Brother Erasmus took full advantage of this – in what seems to have been almost a Darwin tradition, he was not overly healthy. Notwithstanding, neither son was raised to be a member of the idle rich. Both boys went (as boarders) to Shrewsbury School, what is misleadingly known in England as a "public" school, but which is really private. Charles Darwin was later one of the greatest of scientists, and this was anticipated by the two brothers. Even as teenagers, they were messing around with chemicals and running experiments – almost expected of the offspring of a family whose fortune came from the use of chemicals and the like (in making the glazes for pottery). School fare was very different. It consisted of a steady diet of Latin and Greek, interspersed with a little Euclidean geometry. "Much attention was paid to learning by heart the lessons of the previous day; this I could effect with great facility learning forty or fifty lines of Virgil or Homer, whilst I was in morning chapel" (Darwin 1958, 24). *Plus ça change, plus c'est la même chose.* Similar schooling one hundred and fifty years later means that the opening lines of Virgil's *Aeneid – Arma virumque cano* ("I sing of arms and the man") – are engraved on my heart.

Darwin was packed off to Edinburgh to train as a physician, but, horrified by the operations and generally bored with the topic, dropped out after two years. His father then redirected him to Cambridge University, cynically (given his own beliefs) to prepare him for ordination in the Anglican Church. Being a Church of England vicar had the dual advantages of being very respectable and, so long as one had the funds to pay for a curate to do the hard work, quite undemanding. As at school, the official offerings were hardly inspiring, but at Edinburgh Darwin had started to show an obsessive interest in natural history. He continued this at Cambridge – combined with fanatical beetle collecting! At that time (Darwin was there from 1828 to 1831), there were no honors degrees in the sciences (they did not come until 1850), but there were professors in the sciences – usually honorific – and they gave lectures in the sciences (for a fee, not part of the curriculum). Somewhat naturally, those interested in the life sciences (including geology) used to meet. These included Adam Sedgwick, professor of geology, John Henslow, professor of botany, and William Whewell, who was then professor of mineralogy and later professor of philosophy, and author of the *History of the Inductive Sciences* (1837) and the *Philosophy of the Inductive Sciences* (1840). All were, necessarily, ordained members of the Anglican Church. Darwin got into the group and clearly impressed his elders. When he graduated, thanks to the influence of Henslow, Darwin was offered and accepted a position on HMS *Beagle*, about to set off for a trip around South America, mapping the coast. South America was a major trading connection with Britain and so the merchant ships needed good charts. Darwin's official position was unpaid companion to the skipper, Robert Fitzroy, but he soon became de facto ship's naturalist.

The trip lasted from 1831 to 1836, eventually going all the way around the globe (Figure 2.2). When he left England, Darwin was a Christian, amusing the sailors with his literal understanding of the Bible. However, during the voyage, the major scientific interest and influence was the "uniformitarian" theory of geology, being expounded in his *Principles of Geology* (1830–33) by the Scottish lawyer Charles Lyell. Uniformitarianism, a position to which Darwin at once subscribed, argues that all geological phenomena

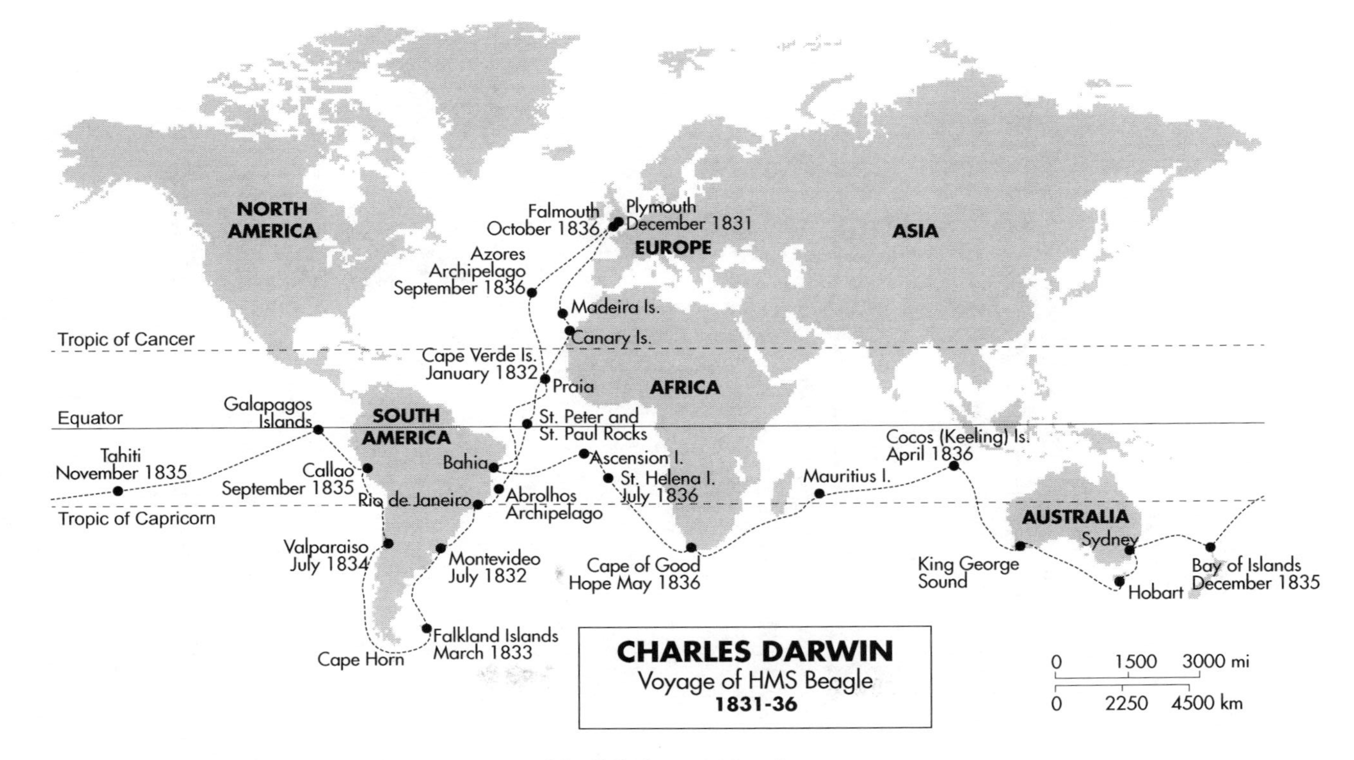

Figure 2.2 Map of the five-year circumnavigation of the globe by HMS *Beagle*.

are produced slowly, over eons of time, by unbroken laws. It is to be contrasted with "catastrophism" – accepted by someone like Whewell – who saw earth's history as a number of stages, possibly (probably) fueled by direct divine intervention. Increasingly, as a function of his uniformitarianism, Darwin started to doubt the literal truth of the biblical miracles, and soon, using the distinction mentioned in Chapter 1, moved from theism (belief in an intervening God) to deism (belief in a God who set the laws in motion and then stood back). This was not a particularly scandalous move to make. Unitarians are deists. This theological position lasted right through the writing of the *Origin*. Later, in around 1865, like many of his contemporaries, Darwin moved to a fairly conventional agnosticism. He was never an atheist and would have considered the New Atheists rather vulgar. Darwin always avoided science–religion conflicts and indeed, after his marriage, his closest personal friend was the local vicar. That they were both gentlemen far outweighed any religious differences.

When Darwin set off, and when he returned, although he knew the basic ideas of the theory from reading his grandfather's works and, when at Edinburgh, interacting with a then-resident physician/naturalist, Robert Grant, who was an evolutionist, he himself was not an evolutionist. That Darwin did not then make the move to evolution is hardly remarkable. At Cambridge, for three years, he would have been subject to countless warnings against the dreadful doctrine. More immediately, on the *Beagle* voyage, he would have thought of himself, a follower of Lyell, as more a geologist than a biologist. In the second volume of his *Principles* (Darwin took the first volume with him and had later volumes shipped out), Lyell included a detailed exposition of Lamarck's theory, which he then critiqued. Uniformitarian though he may have been, Lyell always found the thought of a nonmiraculous start to humankind too big a pill to swallow. As part of his campaign against such a natural origin for humankind, Lyell was keen to stress the nonprogressive nature of the fossil record. He did not want anyone seduced by a record leading from simple to complex, from monad to man. Lyell claimed there was no "foundation in fact" for the belief that "in the successive groups of strata, from the oldest to the most recent, there is a progressive development of organic life, from the simplest to the most complicated forms" (Bartholomew 1973, 281). So here was another factor pointing Darwin away from evolutionary speculations.

Darwin was a bad sailor. When possible, the *Beagle* would be left to its charting while he went overland, looking at the local geology and seeking out phenomena that supported Lyell. Evolution was not on the menu. But other things were. In particular, Darwin wanted to confirm Lyell's "grand new theory of climate" (Lyell 1881, 1, 261). It was clear from fossil evidence that parts of the world had been warmer in the past. Paris had remains of palm trees. Catastrophists explained this as a result of the Earth cooling progressively as it prepared itself for the arrival of humans. Lyell could not ignore this evidence of different climates in the past, so he argued that the world is like a giant waterbed and, as some parts sink, perhaps owing to deposition thanks to rain and rivers and the like, other parts rise. This alters the geography of the continents and, as this happens, things such as the Gulf Stream – which keeps Britain artificially warm – come and go, altering the climate. Thus, the palm trees can be explained without assuming an overall constant cooling. Darwin was convinced, from his study of the geology of South America, that Lyell's theory was correct.

It was while working within this theory, trying to confirm it and use it to explain new phenomena, that Darwin made his first real scientific discovery. Coral reefs – Darwin saw them as the *Beagle* circumnavigated the globe – are those circles of coral in the sea that contain glassy lagoons, often the habitats for a variety of organisms. What causes them? Coral can only survive and grow at the surface of the sea. Lyell supposed that the reefs were the tops of volcanoes with coral growing around their rims. This is improbable, because it supposes that volcanoes constantly reach almost to the top, but not quite. Darwin countered by suggesting that coral grew on the tops of underwater mountains and, as the seabed sank ever lower (in waterbed fashion), the coral keeps growing upwards, just breaking the surface where it thrives (Darwin 1842) (Figure 2.3).

Coral reefs were but one part of the influence and importance of Lyell's theorizing. In trying to work out past history, looking for evidence of the land rising or falling, the geographical distribution of organisms was an important tool. There might not be progress in the record, but there are connections between the nature of organisms and their adaptation to circumstances. Minor changes of circumstance, generally taking less time, demand less change in the organisms. Major changes, taking more time, demand more changes in the organisms. If you find two populations on different sides of a canyon, then, if they are similar, you can infer that the canyon is not that old. If different, then the canyon is aged. Exactly how these changes occur – how organisms appear in the first place and the order in which they appear – Lyell left (suspiciously) blank, although they were supposed to be natural except in the case of humans. At least, one presumes they were supposed to be natural. Lyell is not exactly categorical on this subject: "Each species may have had its origin in a single pair, or individual, where an individual was sufficient, and species may have been created in succession at such times and in such places as to enable them to multiply and endure for an appointed period,

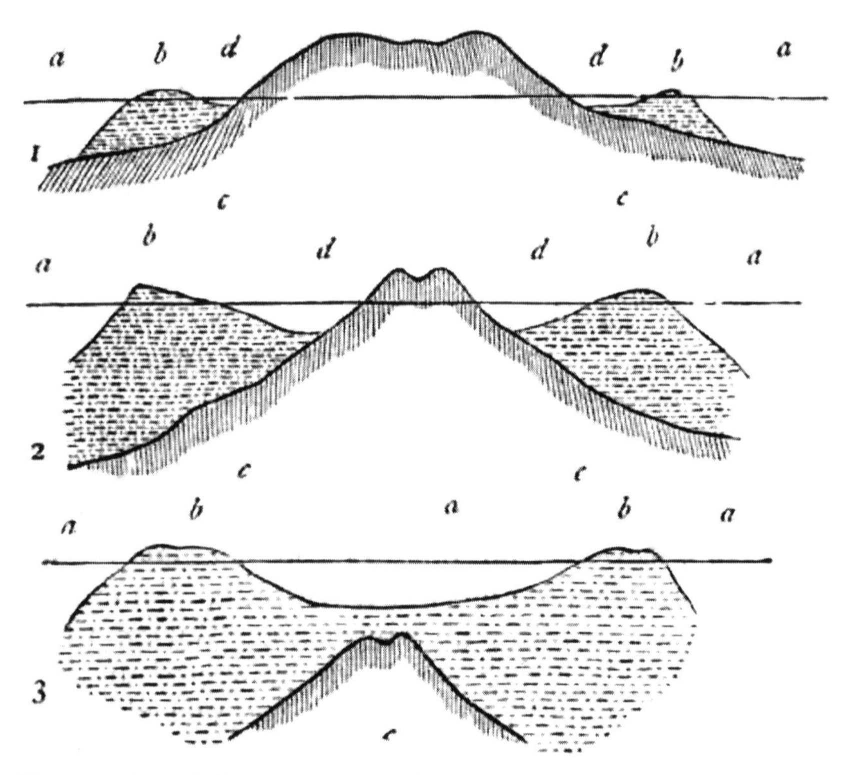

Figure 2.3 Darwin's diagram showing the formation of coral reefs. As the land subsides (1, 2, 3), the coral keeps growing upwards to stay at the sea's surface. Eventually, it is only the coral at the sea's surface.

and occupy an appointed space on the globe" (Lyell 1830–3, 2, 124). To quote a sympathetic historian, "the reader is left entirely in suspense" (Wilson 1972, 339), although the same historian insists that new species are "being produced in some way unknown but as a consequence of the ordinary processes of nature" (Wilson 1971, 48). The important point is that Lyell presupposed that similar ecological niches would call for similar species. So if one found similar species close together, one can infer climate and so forth were unchanged. Different species meant change of climate and the like, and presumably the time span for the first was less than the time span for the second.

Darwin worked assiduously on these kinds of problems. Particularly noteworthy was a visit to the Galapagos Archipelago, a group of islands in the Pacific about a thousand kilometers from the South American mainland (Ecuador) (Figure 2.4). Darwin discovered – he was told – that the denizens

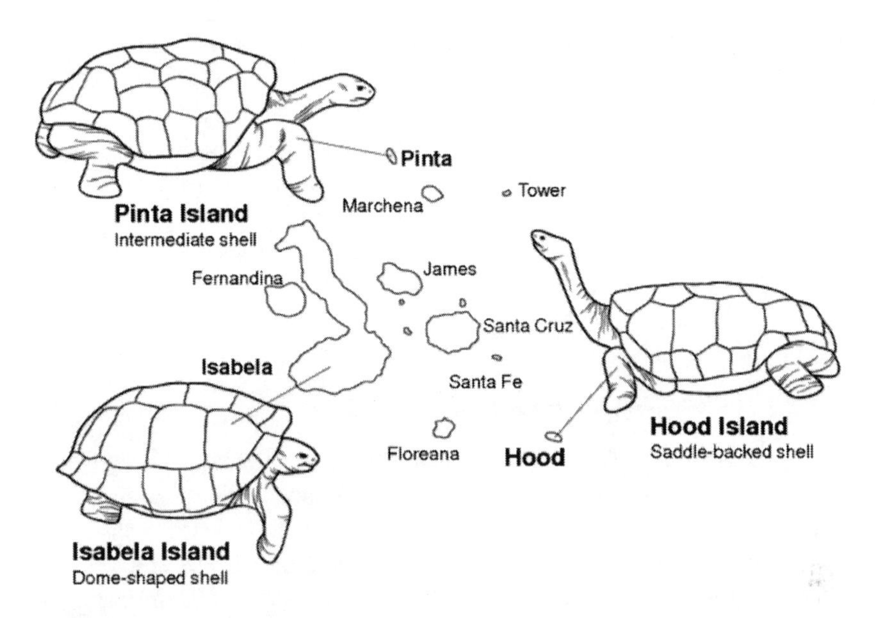

Figure 2.4 Three different species of tortoise, found on three separate islands of the Galapagos.

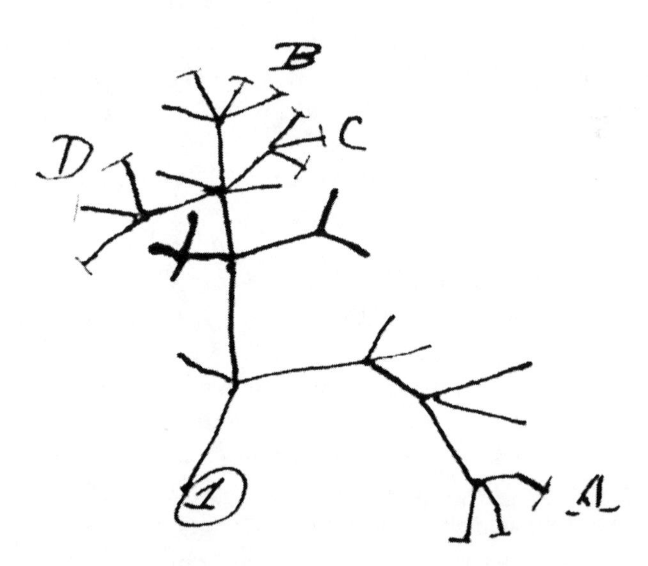

Figure 2.5 Darwin's first sketch of the Tree of Life, *c.* July 1837. Note that this is not a progressive picture with humans at the top.

(birds and reptiles, notably tortoises) of different islands, often within sight of each other, were similar but different. The same was true comparing these island kinds and those found on the mainland – similar but different. On his return to England, in the fall of 1836, Darwin gave his collections to an expert classifier. In the spring of 1837, he learned that the birds of the Galapagos were definitely different species – it was the mocking birds that counted, not the better-known finches – and that was enough to convert him to evolution (Sulloway 1982). He started to keep notebooks, and here we find his first sketch of a tree of life (Figure 2.5).

Darwin: The Evolutionist

Three things of note. First, why Darwin and not others? He knew about evolution, thanks to reading his grandfather and the influence of Grant, more than most others. One suspects that a major factor was that Darwin was on his own for five years and thus used to thinking for himself. He must have been keenly aware of the inadequacy of Lyell's thinking on these questions: "species may have been created in succession" is hardly a ringing endorsement of a natural process. Second, thinking for yourself is one thing. Thinking for yourself under the spotlight is another matter. The very ambitious Charles Darwin kept his evolutionary thinking private (Darwin 1987). On his return to England, all thoughts of parson-hood had vanished. Thanks especially to letters he had been penning to Henslow (as well as collections sent on ahead), not to mention his earlier connections to Cambridge, he was welcomed into the group of professional geologists and life scientists. It paid off. He was elected to the Athenaeum Club in 1838 and to the Royal Society in 1839 (at the age of thirty). He was also a member of the Geological Society, and Whewell, who had taken on a role as mentor, pushed him into becoming a secretary of the society. He wrote a major paper on the "roads" of Glen Roy that was published by the Royal Society (Darwin 1839). Announcing that you were now an evolutionist would not further your career. (Glen Roy is a valley in Scotland, distinguished by parallel paths running around the sides. Darwin argued, from a Lyellian perspective, that they were caused by the sea, now gone because the land had risen. He was proven wrong by the Swiss naturalist Louis Agassiz, who showed that the missing water had been kept blocked up by glaciers, which then melted and let the water run free.)

Third, what we today would (rightly) deem the most important, Darwin was a fully committed Newtonian, thinking (as did Kant) that the Newtonian mechanics of the *Principia* was the model for any successful science. Darwin got this from the scientists with whom he associated while an undergraduate at Cambridge. Whewell, in particular, who paid most attention to the philosophical aspects of science, was nigh fanatical in his enthusiasm – entirely appropriately for one who was to become the Master of Trinity College, Newton's old college. And what was Newton's great achievement? It was to show how the celestial laws – Copernicus and Kepler – and the terrestrial laws – Galileo – are all to be seen as consequences of one overall, binding cause, the force of gravitational attraction. Something, incidentally, that was part of God's plan. "We must conceive the Deity, not only as constructing the most refined and vast machinery, with which, as we have already seen, the universe is filled; but we must also imagine him as establishing those properties by which such machinery is possible: as giving to the materials of his structure the qualities by which the material is fitted to its use" (William Whewell 1833, 300).

Darwin, thanks to Whewell, would have known of Kant's assessment of biology:

> It is indeed quite certain that we cannot adequately cognise, much less explain, organised beings and their internal possibility, according to mere mechanical principles of nature; and we can say boldly it is alike certain that it is absurd for men to make any such attempt or to hope that another *Newton* will arise in the future, who shall make comprehensible by us the production of a blade of grass according to natural laws which no design has ordered. We must absolutely deny this insight to men. (Kant 1790)

An eager young man like Darwin would have seen that, having become an evolutionist, the next task was to find the biological equivalent of gravitation. Watch out blades of grass. Here comes Charles Darwin, the would-be Newton of biology! He started a series of very private notebooks in which he jotted his thinking (Darwin 1987). Darwin knew of Lamarckism, and indeed adopted it as a cause and never relinquished that view. But he saw it could not really do the whole job. "Wax of Ear, bitter perhaps to prevent insects lodging there, now these exquisite adaptations can hardly be accounted for by my method of breeding there must be some car[r]elation, but the whole mechanism is so beautiful" (Notebook C 174). This was written before the discovery of natural selection, when his only "method of breeding" was the inheritance of acquired characteristics. Darwin soon twigged that selective breeding was a major clue to what happens in the wild. Indeed, he even read a pamphlet that told him explicitly about the analogy between the world of the breeder and the harsh world of nature. Sir John Sebright wrote:

> A severe winter, or a scarcity of food, by destroying the weak or unhealthy, has all the good effects of the most skilful selection. In cold and barren

> countries no animal can live to the age of maturity, but those who have strong constitutions; the weak and the unhealthy do not live to propagate their infirmities, as is too often the case with our domestic animals. To this I attribute the peculiar hardiness of the horses, cattle, and sheep, bred in mountainous countries, more than their having been inured to the severity of climate… (Sebright 1809, 15–16; Ruse 1975b)

Darwin took careful note of this passage, and, even though he could not quite see the full import, grasped that if something like this went on long enough, we would get full-blooded species. "Sir J. Sebright—pamphlet most important showing effects of peculiarities being long in blood.++ thinks difficulty in crossing race—bad effects of incestuous intercourse.—excellent observations of sickly offspring being cut off so that not propagated by nature.—Whole art of making varieties may be inferred from facts stated.—" (C, 133). (I am not accusing Darwin of plagiarism. Sebright had no thoughts of evolution.)

Then, in the early fall of 1838, Darwin read the sixth edition (1826) of the work *An Essay on the Principle of Population* (first edition 1798) by the Anglican clergyman (Thomas) Robert Malthus. Thanks to the Industrial Revolution, there was a huge population explosion. In rural England, there was a big incentive to keep families reasonably small. Apart from anything else, one often had to wait until one inherited property even to start a family. In urban England, large families were an asset. Many factory jobs (such as getting under the loom to repair broken threads) were filled by children. Arguing that this increase was no anomaly, Malthus argued that population numbers would always outstrip resources:

> If it be allowed that by the best possible policy the average produce could be doubled in the first 25 years, it will be allowing a greater increase than could with reason be expected. In the next 25 years it is impossible to suppose that the produce could be quadrupled. It would be contrary to our knowledge of the properties of land.

Continuing that, even if we could double production every twenty-five years, we would still be in trouble. Population numbers increase geometrically; food and space supplies at the most arithmetically:

> Taking the whole earth, the human species would increase as the numbers 1, 2, 4, 8, 16, 32, 64, 128, 256, and subsistence as 1, 2, 3, 4, 5, 6, 7, 8, 9. In two centuries the population would be to the means of subsistence as 256 to 9; in three centuries as 4,096 to 13, and in two thousand years the difference would be almost incalculable. (Malthus 1826, I, 1)

This all leads to what Malthus christened a "struggle for existence."

Darwin seized on this. He had the motive force behind an ongoing "natural selection." Quoting from a private notebook dated September 28, 1838:

> Population is increase at geometrical ratio in far shorter time than 25 years – yet until the one sentence of Malthus ["It may safely be pronounced, therefore, that the population, when unchecked, goes on doubling itself every twenty five years, or increases in a geometrical ratio."] no one clearly perceived the great check amongst men. – there is spring, like food used for other purposes as wheat for making brandy. – Even a few years plenty, makes population in Men increase & an ordinary crop causes a dearth. take Europe on an average every species must have same number killed year with year by hawks, by cold &c. – even one species of hawk decreasing in number must affect instantaneously all the rest. – The final cause of all this wedging, must be to sort out proper structure, & adapt it to changes. – to do that for form, which Malthus shows is the final effect (by means however of volition) of this populousness on the energy of man. One may say there is a force like a hundred thousand wedges trying force into every kind of adapted structure into the gaps of in the oeconomy of nature, or rather forming gaps by thrusting out weaker ones. (D 135e)

It is not long before we find Darwin making explicit reference to his new mechanism. Interestingly, showing he had none of the hangups of Lyell about natural human origins, it is about our species: "An habitual action must some way affect the brain in a manner which can be transmitted.—this is analogous to a blacksmith having children with strong arms.—The other principle of those children, which chance? produced with strong arms, outliving the weaker ones, may be applicable to the formation of instincts, independently of habits" (N 42).

The next major move in Darwin's evolutionary thinking came just over three years later, when he wrote up his ideas more formally in a short "Sketch" (as it is now called), and then two years later in a longer "Essay." There are several understandable reasons for the delay. Darwin was finishing up his contribution to the official report on the *Beagle* voyage (*Journal of Researches*). Later, it was republished in popular form, to become known as *The Voyage of the Beagle*, and – given the avid readership of accounts of geographical expeditions – this firmly located Darwin in the public eye. Then, early in 1839, he married his first cousin Emma Wedgwood. Not only was this union much appreciated by both families – no outsider would get their hands on the money – but it also proved to be a very happy marriage, lasting over forty years until Darwin's death in 1882. They bought a house in the village of Downe in Kent, moving there and remaining. Giving proof of their huge, combined wealth, as soon as they paid cash for the house, they set about building an extension, anticipating their large family which, wasting no time, started with the birth of William at the end of 1839.

Then, the abiding mystery. Darwin started to show symptoms of an illness that was to plague him all his adult life – nausea, sickness, headaches, boils, depression, flatulence (lots), and more. There have been numerous hypotheses about its cause. Psychological factors, such as a fear of publishing his evolutionary ideas, are generally now discounted. Darwin knew he had won and never lost his confidence in that. Long popular was the suggestion that he suffered from Chaga's disease, a function of a parasite picked up in South America. Today, the most favored explanation is lactose intolerance (Dixon and Radick 2009). We know that when he went off to health spas, where he lived on gruel, he felt much better. We know also that when he returned home, he started to feel sick again. We have Emma Darwin's cookbook, and the recipes involve huge amounts of milk and cream. How about ending a meal with "Lady Skymaston's Pudding"?

> 5 eggs to a pint of cream –
> The cream must be boiled
> & made very sweet
> & poured upon the eggs:
> 2 ounces of sugar
> boiled with a little water till it candies
> & put in the bottom of the mould:
> let the cream stand till it is quite cold
> & then pour it in the mould –
> Steam it 20 minutes before turning out. –

After a helping of this, I doubt you need lactose intolerance to feel nauseous.

Darwin did not make an absolute secret of his evolutionary inclinations. It was around this time that he became friends with the botanist Joseph Dalton Hooker (1817–1911), and the younger man was soon let into the secret:

> At last gleams of light have come, & I am almost convinced (quite contrary to opinion I started with) that species are not (it is like confessing a murder) immutable. Heaven forfend me from Lamarck nonsense of a "tendency to progression" "adaptations from the slow willing of animals" &c, – but the conclusions I am led to are not widely different from his – though the means of change are wholly so – I think I have found out (here's presumption!) the simple way by which species become exquisitely adapted to various ends. (Letter to Hooker, January 11, 1844)

Note that Darwin is pushing his theory away from thoughts of progress, undoubtedly because of the influence of Lyell, but also, as we shall see, because from the first he could see that natural selection does not support progress.

From then on, there is a lot of correspondence on the topic, especially after Darwin let Hooker read his 1844 *Essay*. Hooker offered constructive criticisms, although, given Darwin's illness, it was hard for them to have any detailed face-to-face discussion. However, when it came to the public world, Darwin remained silent. It was not, absolutely not, because he felt his work unfinished. There is one additional major addition he was to make in the 1850s, and there were other extensions and clarifications. Expectedly, in both *Sketch* and *Essay*, Darwin starts his discussion based on the order of his discovery, meaning he starts with human breeding – "if man selects, then new races rapidly formed" – and by the time of the *Origin* Darwin had spent much time delving into the mysteries of pigeon breeding, material of which he makes much use in the version he finally published. But what is striking is how the whole structure and the argumentation of the *Origin* is to be found outlined in the *Sketch* and significantly extended in the *Essay*. In the *Sketch* (repeated in the *Essay*), he even has, virtually word for word, the final lines of the *Origin*:

> There is a simple grandeur in the view of life with its powers of growth, assimilation and reproduction, being originally breathed into matter under one or a few forms, and that whilst this our planet has gone circling on according to fixed laws, and land and water, in a cycle of change, have gone on replacing each other, that from so simple an origin, through the process of gradual selection of infinitesimal changes, endless forms most beautiful and most wonderful have been evolved. (F. Darwin 1909, 52)

Darwin left money for the publication of his *Essay* should he die – he wanted the glory of his discovery, but not just now.

Vestiges

Why? The reason is that, in 1844, Robert Chambers, a very successful Scottish businessman, with his brother William responsible for the popular weekly *Chambers's Edinburgh Journal*, anonymously published his evolutionary tract *Vestiges of the Natural History of Creation*. There was no natural selection, but the message of evolution – a thoroughly progressivist evolution – was there in full force:

> The idea, then, which I form of the progress of organic life upon the globe—and the hypothesis is applicable to all similar theatres of vital being—is, *that the simplest and most primitive type, under a law to which that of like-production is subordinate, gave birth to the type next above it, that this again produced the next higher, and so on to the very highest*. (Chambers 1844, 222, his italics)

The discussion starts with the "nebular hypothesis," a suggestion by Immanuel Kant, refined by Pierre Laplace, that the universe formed in a kind of evolutionary pattern. Then, turning to the organic world here on our planet, the anonymous author starts with the formation of life, and then takes us up the ladder of life until we arrive at the top, human beings. Rhetorically, it asked: "That God created animated beings, as well as the terraqueous theatre of their being, is a fact so powerfully evidenced, and so universally received, that I at once take it for granted. But in the particulars of this so highly supported idea, we surely here see cause for some re-consideration" (152). It is ridiculous to suggest that God created again and again to produce this progressive stream of life: "Some other idea must then be come to with regard to the mode in which the Divine Author proceeded in the organic creation." What could this be other than, as with the nebular hypothesis, God must have created through unbroken law. Evolution. And turning the hose towards the critics: "When all is seen to be the result of law, the idea of an Almighty Author becomes irresistible, for the creation of a law for an endless series of phenomena—an act of intelligence above all else that we can conceive—could have no other imaginable source" (153).

There was frenzied reaction to this vile work – reaction from Darwin's group including Sedgwick and Whewell. The latter could not really be bothered to counter *Vestiges*, so simply put together a collect of abstracts from earlier writings – *Indications of a Creator* (1845) – with a spirited preface showing the impossibility of evolution. Humans have powers of reason totally missing in animals. "There are, in animals, no germs of this power of abstraction, this apprehension of abstract and general Truth. The Instinct of animals cannot become the Reason of man, by any process of developement [*sic*]. We cannot unfold the mind of a spider or a bee into the mind of a geometer" (xiii). We know that Darwin was worried about this sort of thing, for he wrote to Lyell that he had read Sedgwick's review:

> though I find it is far from popular with non-scientific readers. I think some few passages savour of the dogmatism of the pulpit, rather than of the philosophy of the Professor chair; & some of the wit strikes me as only worthy of Broderie in the Quarterly. Nevertheless it a grand piece of argument against mutability of species; & I read it with fear & trembling. (Letter to Lyell, October 8, 1845)

He assured Lyell, although one suspects he was more intent on assuring himself, "that I had not overlooked any of the arguments, though I had put them to myself as feebly as milk & water." Hardly the most confident of emotions.

In a way, Darwin's complete acceptance by and immersion in the professional science of the day rather backfired. Already there was a feeling that Darwin was not entirely orthodox on the species question and there were suspicions that he might be the anonymous author of *Vestiges*. He wrote to a Cambridge undergraduate friend: "Have you read that strange unphilosophical, but capitally-written book, the Vestiges, it has made more talk than any work of late, & has been by some attributed to me.—at which I ought to be much flattered & unflattered" (Letter to W. D. Fox, April 24, 1845). Keeping clear of controversy, Darwin began what became an eight-year study of barnacle taxonomy, a safe occupation. To be fair, Darwin's primary motive was that the very critical reception of *Vestiges* showed him that he really had to be on top of his biology, and sweeping unjustified claims about variation would not do. His study of barnacles, which surely lasted longer than he anticipated – bad health was a major factor – filled the gap (Darwin 1851a, b, 1854a, b).

Only in the early 1850s, when this study was completed, did Darwin turn again to evolution. By this time, the older generation was starting to fade away and more secular newcomers were appearing on the scene. In addition, evolution was more talked about. Some comments showed that, despite the criticism, *Vestiges* was not entirely without effect. The major poet of the day, Alfred Tennyson, picked up on Chambers's ideas in his poem *In Memoriam* (1850), a tribute to a dead friend of his youth. Depressed by the seemingly meaninglessness of it all – it was he who popularized the famous phrase about "nature red in tooth and claw" – Tennyson found optimistic relief in Chambers's thinking, suggesting that perhaps this dead friend was a special superior being who came before his time. A positive answer to a question posed in *Vestiges*: "Is our race but the initial of the grand crowning type?" (276).

> And, moved thro' life of lower phase,
> Result in man, be born and think,
> And act and love, a closer link
> Betwixt us and the crowning race.

Thus encouraged – threatened? – Darwin started to write again on evolution, working on a massive book that would answer everything, in detail and at length (Darwin 1975). Fortunately, for those of us who think the reading of books of this size should be left to graduate students looking for material for their dissertations, in June 1858 Darwin's work was interrupted by the arrival of an essay from a younger naturalist, Alfred Russel Wallace (1858), out in the East Indies – an essay which, independently, mirrored Darwin's ideas. Through his friends Hooker and Lyell, Darwin arranged for the immediate publication of Wallace's essay, together with extracts from his own earlier writings. Fifteen months later, he had written *On the Origin of Species by Means of Natural Selection, or the Preservation of Favoured Races in the Struggle for Life.* It was published on November 24, 1859. It was from the first a strong seller, and was to go through six, much rewritten, editions, the last appearing in 1872.

3

The Origin of Species

The "Origin"

The *Origin* is deceptive. It is written in a user-friendly style. No mathematics for a start! It is, however, very carefully structured (Ruse 1975a; Figure 3.1). Darwin knew fully what he was about. He avoided entirely discussion about the beginning of life, presumably naturally from the inorganic. This was exactly the time when Louis Pasteur was driving the knife into the heart of spontaneous generation (Farley 1977). Darwin knew that that was a bog from which it would be impossible to escape. He refers to the "original parent," but its nature and genesis is left as an exercise for the reader (Darwin 1859, 121).

What Darwin did do was offer us a work guided by what the philosophers of science of his day said about quality science. Two figures stand out. First, John F. W. Herschel, an astronomer, like his father before him, and the author of a popular book on the nature of science – *The Preliminary Discourse on the Study of Natural Philosophy* – published in 1830. It was recommended to Darwin by Whewell, and he read it just before the *Beagle* voyage and spoke of it as one of the most influential books he had ever read. The second was Whewell himself, who expressed his ideas most fully in his *History of the Inductive Sciences* (1837), read twice by Darwin in the year it appeared, and his *Philosophy of the Inductive Sciences* (1840), a very detailed and accurate review of which, by Herschel (1841), appeared just after it was published and which was studied carefully by Darwin. What makes for an interesting and significant contrast is that both philosophers were ardent Newtonians, thinking that good science is (what we today call) hypothetico-deductive, like the *Principia*, with causal laws at the beginning and then empirical laws deduced from them. However, when it came to identifying and justifying the central cause, what they both followed

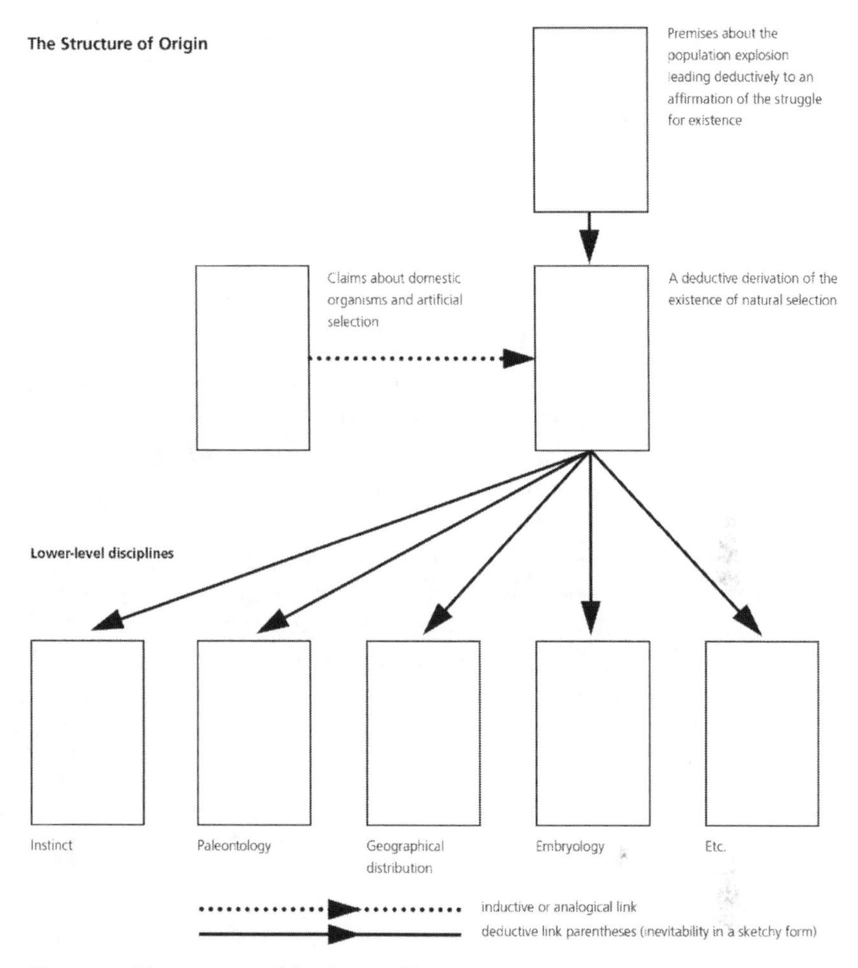

Figure 3.1 The structure of the *Origin of Species*.

Newton in calling a "*vera causa*" ("true cause"), Herschel and Whewell took different paths (Kavaloski 1974).

For Herschel, an empiricist, analogy is the key. We believe in a *vera causa* if we have physical experience of something similar. "If the analogy of two phenomena be very close and striking, while, at the same time, the cause of one is very obvious, it becomes scarcely possible to refuse to admit the action of an analogous cause in the other, though not so obvious in itself" (Herschel 1830, 142). Herschel gives an example:

> For instance, when we see a stone whirled round in a sling, describing a circular orbit round the hand, keeping the string stretched, and flying away the

> moment it breaks, we never hesitate to regard it as retained in its orbit by the tension of the string, that is, by *a force* directed to the centre; for we feel that we do really exert such a force. We have here *the direct perception* of the cause. When, therefore, we see a great body like the moon circulating round the earth and not flying off, we cannot help believing it to be prevented from so doing, not indeed by a material tie, but by that which operates in the other case through the intermedium of the string, – a *force* directed constantly to the centre. (Herschel 1830, 142)

Lyell's theory of climate is instanced as a perfect exemplar of such an empirically based cause. "A cause, possessing the essential requisites of a *vera causa*, has ... been brought forward, in the varying influence of the distribution of land and sea over the surface of the globe" (139).

Whewell was a rationalist. For him, it was the consequences of a causal statement that counted, especially if the consequences led to different areas of study, thus brought together by the cause: what Whewell (1840) called a "consilience of inductions" (2, 230). Thanks to innovative demonstrations, such as the interference patterns yielded by Young's double-slit experiment, the wave theory of light (championed in the seventeenth century by Christian Huygens) had recently displaced the particulate theory of light (championed by Isaac Newton). Writing of the wave theory as the "undulatory" theory, Whewell wrote of

> the evidence in favour of the undulatory theory of light, when the assumption of the length of an undulation, to which we are led by the colours of thin plates, is found to be identical with that length which explains the phenomena of diffraction; or when the hypothesis of transverse vibrations, suggested by the facts of polarization, explains also the laws of double refraction. When such a convergence of two trains of induction points to the same spot, we can no longer suspect that we are wrong. Such an accumulation of proof really persuades us that we have to do with a vera causa. (2, 447)

Darwin's genius was to appeal to both the empiricist criterion for *vera causa* and the rational criterion also, both sides to the argument backing the hypothetico-deductive picture that incorporates Darwin's cause, natural selection:

> In fact the belief in natural selection must be at present grounded entirely on general considerations. (1) on its being a vera causa, from the struggle for existence; & the certain geological fact that species do somehow change (2) from the analogy of change under domestication by man's selection. (3) & chiefly from this view connecting under an intelligible point of view a host of facts. (Letter to George Bentham, May 22, 1863)

Human-Caused Selection

Gallia est omnis divisa in partes tres. The *Origin* likewise is divided into three parts. The first part – "Chapter I: Variation under Domestication" – follows Herschel's directives. It deals with artificial selection, setting it up as a perceived analogy to the inferred natural selection. To this end, Darwin discussed in some detail the successes of breeders, both in the farmyard – cows, pigs, and the like – and in the world of fanciers, pigeon breeders particularly. "Altogether at least a score of pigeons might be chosen, which if shown to an ornithologist, and he were told that they were wild birds, would certainly, I think, be ranked by him as well-defined species" (22). They might well be considered members of different genera. However: "Great as the differences are between the breeds of pigeons, I am fully convinced that the common opinion of naturalists is correct, namely, that all have descended from the rock-pigeon (Columba livia)..." (23).

How does this come about? "The key is man's power of accumulative selection: nature gives successive variations; man adds them up in certain directions useful to him. In this sense he may be said to make for himself useful breeds" (30). One interesting topic introduced at the end of this first chapter of the *Origin* is "unconscious selection." Referring to an opinion of William Youatt, expert on animal breeds, Darwin wrote of two flocks of Leicester sheep (owned by a Mr. Buckley and a Mr. Burgess) that after separation had, within fifty years, become very different in features. There was no conscious intention driving this change, "and yet the difference between the sheep possessed by these two gentlemen is so great that they have the appearance of being quite different varieties" (36). Darwin drew our attention to this phenomenon for the obvious reason that here we have selection, although human-caused, taking a direction altogether unplanned by humans. An even closer analogy to natural selection, which is the topic of the second part of Darwin's *Origin.*

Natural Selection

This section (Chapters II, III, and IV), the hypothetico-deductive part, starts with the claim that in all populations one is going to find variations. "No one supposes that all the individuals of the same species are cast in the very same mould. These individual differences are highly important for us, as they afford materials for natural selection to accumulate, in the same manner as man can accumulate in any given direction individual differences in his domesticated productions" (45). Darwin had always believed this, but, by the time of the *Origin*, his beliefs were backed by his eight-year study of barnacles. The important point is not that the variations are uncaused – Darwin the Newtonian would never claim this – but that they are "random" in the sense of not occurring for need and probably, usually, on appearance, of no great benefit to their possessors.

Now Darwin was ready to move to his Newtonian-like cause. First, there is the "struggle for existence":

> A struggle for existence inevitably follows from the high rate at which all organic beings tend to increase. Every being, which during its natural lifetime produces several eggs or seeds, must suffer destruction during some period of its life, and during some season or occasional year, otherwise, on the principle of geometrical increase, its numbers would quickly become so inordinately great that no country could support the product. Hence, as more individuals are produced than can possibly survive, there must in every case be a struggle for existence, either one individual with another of the same species, or with the individuals of distinct species, or with the physical conditions of life. (63–64)

Then, combining the struggle with the variation in populations, we move on to "natural selection":

> Let it be borne in mind in what an endless number of strange peculiarities our domestic productions, and, in a lesser degree, those under nature, vary; and how strong the hereditary tendency is. Under domestication, it may be truly said that the whole organisation becomes in some degree plastic. Let it be borne in mind how infinitely complex and close-fitting are the mutual relations of all organic beings to each other and to their physical conditions of life. Can it, then, be thought improbable, seeing that variations useful to man have undoubtedly occurred, that other variations useful in some way to each being in the great and complex battle of life, should sometimes occur in the course of thousands of generations? If such do occur, can we doubt (remembering that many more individuals are born than can possibly survive) that individuals having any advantage, however slight, over others, would have the best chance of surviving and of procreating their kind? On the other hand, we may feel sure that any variation in the least degree injurious would be rigidly destroyed. This preservation of favourable variations and the rejection of injurious variations, I call Natural Selection. (80–81)

Unpacking

A number of points before we move on. First, and most importantly, Darwin thought he had solved the problem of final cause – the problem being that of explaining final cause in mechanical terms. Note "explaining" – not "explaining away" (Lennox 1993). Darwin was always a teleologist, and through the *Origin* quite unselfconsciously referred to final causes:

> It is now commonly admitted that the more immediate and *final cause* of the cuckoo's instinct is, that she lays her eggs, not daily, but at intervals of two or three days; so that, if she were to make her own nest and sit on her own eggs, those first laid would have to be left for some time unincubated, or there would be eggs and young birds of different ages in the same nest. (216–17, my italics).

Adding that: "If this were the case, the process of laying and hatching might be inconveniently long, more especially as she has to migrate at a very early period; and the first hatched young would probably have to be fed by the male alone."

"Taking randomness [of variations] as its starting point, the *Origin*'s tour-de-force is in managing to recoup all results of classical teleology" (Hoquet 2018, 113). The key to such thinking, why Darwin has a claim to be the Newton of biology and why one can think of Darwin as completing the Scientific Revolution, lies in the fact that selection leads to *adaptations*, the features organisms have to survive and reproduce. "How have all those exquisite adaptations of one part of the organisation to another part, and to the conditions of life, and of one distinct organic being to another being, been perfected?" (60). Because features have in the past enabled organisms to survive and reproduce, we project to the future assuming that they will continue to enable. We could be wrong. Climatic change or a new predator might mean that features formerly helpful are now not helpful. But that is always the risk with final-cause explanation: the missing goal object. We have seen that this is not a new problem with Darwin. For Plato, final causes escape from the missing goal problem because we have the idea of the desired end. For Aristotle, final causes escape the problem because the vital force is directed towards the desired end. For Darwin, final causes escape the problem because that is what happened in the past. If things change so the intended final cause never materializes, it is still the case that the final causes occurred in the past, and it is an inductively reasonable supposition that they will go on occurring in the future. Value is imputed not discovered. There is no Platonic Designer "out there." There are no Aristotelian forces "out there." Final causes without tears, but still very much final causes. Darwin is a mechanist. Darwin is also a full-blooded teleologist. Kant has been answered.

Note however, agreeing with Plato and Aristotle about the teleological nature of individual organisms did not imply that Darwin agreed with the philosophers about the teleological nature of the history of life, with humans at the top. There is little doubt that Darwin thought humans were the top and that certain islanders living off the coast of Europe – "This precious stone set in the silver sea" – were the top of the top, but that was not an implication of his theory. From the first, Darwin saw that the natural selection of random variations rules out teleological direction to the top. What works is what works. Again, there are no value forces "out there." What Darwin himself thought and what Darwin thought his theory implied were two different things, even though there were times when he had trouble keeping the two apart. More on this as we go along.

Second, Darwin introduced a secondary mechanism, sexual selection (Richards 2017). A struggle for mates, not a struggle against the elements. "This depends, not on a struggle for existence, but on a struggle between the males for possession of the females; the result is not death to the unsuccessful competitor, but few or no offspring" (88). Actually, Darwin distinguished male combat – stags fighting for the harem – and female choice – peahens choosing the peacock with the most magnificent tail feathers. The idea of sexual selection is not an afterthought. Right back in the fall of 1838, as soon as he discovered natural selection, Darwin was floating ideas that crystallize into sexual selection. "Is it Male that assumes change, & is the offspring brought back to earlier type by Mother?—do these differences indicate, species changing forms, ~~& loosing do~~ if so domestic animals ought to show them.—Anyhow not connected with habits" (D 147e). Darwin makes little use of this form of selection in the *Origin*. Later, it will be different.

For now – a third point – note that sexual selection points to an important aspect of Darwin's thinking about selection. It is always for the benefit of the individual (including one's relatives) and not the group (Ruse 2022b). From the first, Wallace had thought of selection as something operating at the group level. The title of the 1858 paper flags this: "On the tendency of *varieties* to depart indefinitely from the original type" (my italics). One can infer, therefore, that he supposed that the sterility of hybrids, such as the mule, was a function of selection working in favor of the parental groups – horses and donkeys don't want offspring, literally neither fish nor fowl. Revealingly, Darwin responded:

> Let me first say that no man could have more earnestly wished for the success of N. selection in regard to sterility, than I did; & when I considered a general statement, (as in your last note) I always felt sure it could be worked out, but always failed in detail. The cause being as I believe, that natural selection cannot effect what is not good for the individual, including in this term a social community. (Letter to Wallace, April 6, 1868)

A fourth and final point is that Darwin showed little interest in giving actual empirical evidence of natural selection in action. He was not against gathering empirical evidence, including that involving experiments. In the 1850s, now back to working directly on evolution, he ran some experiments with seeds immersed in salt water, to study how far they might travel from one area of land across water to another such area. However, when it came to natural selection, his examples were all pretend. "In order to make it clear how, as I believe, natural selection acts, I must beg permission to give one or two imaginary illustrations" (90). He then went on to talk about wolves chasing deer. "[I] see no reason to doubt that the swiftest and slimmest wolves would have the best chance of surviving, and so be preserved or selected." Given that this was not a function of the *Origin* being a book produced quickly, under pressure – in his big unfinished work on natural selection, he follows the same pattern of referring only to pretend examples (van Wyhe 2002) – this leads to the somewhat heretical suggestion that Darwin-as-scientist was not much of a Darwin-as-natural-selection booster. This is true. Darwin-as-scientist had spent eight years classifying barnacles, and, by his own admission in the *Origin*, to a classifier, adaptation is a nuisance – what counts is "homology," the shared structures linking together very different seeming organisms. "Nothing can be more hopeless than to attempt to explain this similarity of pattern in members of the same class, by utility or by the doctrine of final causes" (435).

This nigh indifference to real examples of natural selection explains one rather curious fact. Darwin never in print referred to natural selection as a "mechanism" (Ruse 2005), where we are using this in the sense of specific causal process as opposed to the sense of an overall root metaphor. There are over two hundred uses of "mechanism" in his correspondence, to and from Darwin, but never in the context of natural selection as (or as not) a mechanism. (Confusingly, the editors are always using it in this way.) Darwin, in print and in correspondence, reserved this word for actual machine-like entities. As in the next book after the *Origin*, on orchids, the use is frequent and not self-conscious. (We have seen him in a notebook – C174 – talking about the wax in the ear as a "mechanism," but not the Lamarckian process that supposedly produced it.) Why never calling natural selection a mechanism? The answer is simple. When scientists speak of natural selection as a mechanism, they have in mind a specific instance where natural selection is actually linked to a change. Thus, when paleontologists explain the plates on the back of the Stegosaurus in terms of heat convection (the wind cools the brute when it is standing in the sun, foraging), it is appropriate to say that natural selection is the mechanism that brings it all about (Farlow, Thompson, and Rosner 1976). Just as a tractor is the mechanism whereby the farmer plows his fields.

Darwin never linked selection to actual cases of adaptation, so there is no reason to call it a mechanism. By the turn of the century, people had moved on to real examples, so gradually they started to extend the language that pointed to the machine metaphor way of thinking. In what is, candidly, a not entirely sympathetic discussion of natural selection, *Darwinism Today*, Stanford biologist Vernon L. Kellogg tells us that he is after the "factors and mechanism of organic evolution" (Kellogg 1905, iii). He is happy with the language even though he is not happy with the process to which it refers. Three decades later, Theodosius Dobzhansky, in (what we shall learn is his fundamental) *Genetics and the Origin of Species*, tells his reader that "mechanisms that counter-act the mutation pressure are known to exist. Selection is one of them..." (Dobzhansky 1937, 37–38). And down to the present up on the internet so one and all can see: *Evolution 101*: "Natural selection is one of the basic mechanisms of evolution" (Museum of Paleontology 2023, Berkeley).

And On to the Tree of Life

Moving on (to complete the second part of the *Origin*), Darwin made use of a significant insight, the relevance of which occurred to him only in the 1850s. That is, fifteen years after he had discovered natural selection and ten years after he wrote the "Sketch" and the "Essay." He introduced Adam Smith's notion of the "division of labor." We work for our own ends and, thanks to the Divine Scotsman Up Above, everything meshes and is for the good of all. "It is not from the benevolence of the butcher, the brewer, or the baker that we expect our dinner, but from their regard to their own self-interest. We address ourselves not to their humanity but to their self-love, and never talk to them of our own necessities, but of their advantages" (Smith 1776, 18). Picking up on this, Darwin wrote: "No naturalist doubts the advantage of what has been called the 'physiological division of labour'; hence we may believe that it would be advantageous to a plant to produce stamens alone in one flower or on one whole plant, and pistils alone in another flower or on another plant" (93–94).

This is true at the group level also: "The advantage of diversification in the inhabitants of the same region is, in fact, the same as that of the physiological division of labour in the organs of the same individual body." Hence: "in the general economy of any land, the more widely and perfectly the animals and plants are diversified for different habits of life, so will a greater number of individuals be capable of their supporting themselves." Although we are dealing with things at the group level, understand that it is selection working

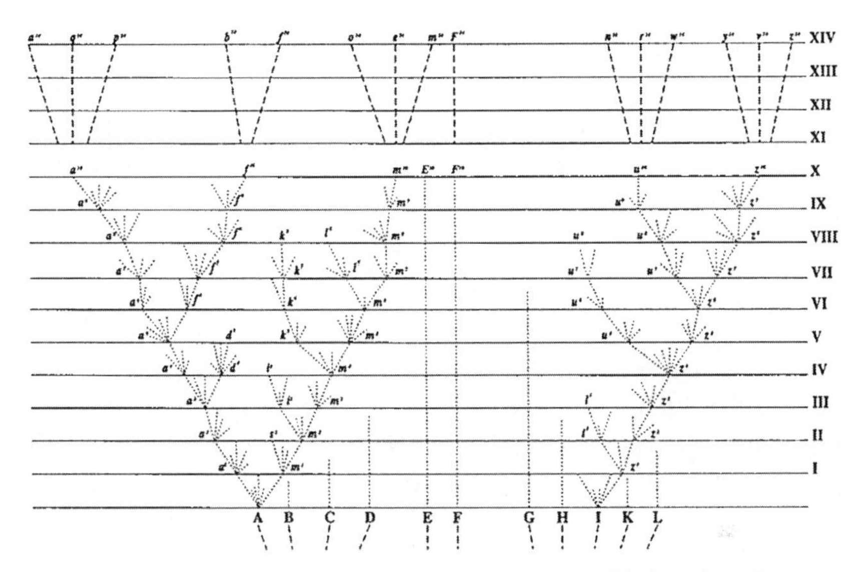

Figure 3.2 Darwin's sole illustration in the *Origin*, (1859), a tree-like branching diagram.

at the individual level that is the causal factor – individuals capable of self-support (115–16).

Finally, Darwin was ready to turn to the overall picture. As Newton's cause explains the overall picture of the heliocentric universe, Darwin's cause explains the overall picture of the history of life. It is like a great tree "which fills with its dead and broken branches the crust of the earth, and covers the surface with its ever branching and beautiful ramifications" (130). Darwin illustrated his claim with the only figure in the *Origin* (Figure 3.2). Note that this figure is intended to show branching and says nothing about the actual course of life. Note also that, without the division of labor making new species branching from older forms, the tree of life goes unexplained. There is no mention of it in either the *Sketch* or the *Essay*.

In this second section, nothing is very formal, but one can see the nature of the argument. We start with premises about variation and the struggle, and we go on to infer natural selection. (A not-very-important break with Newton is that, whereas for the physicist his cause, gravitational attraction, has an axiom of its own, for the life scientist his cause, natural selection, is inferred from earlier axioms. As Darwin said in his letter to Bentham: "on its being a vera causa, from the struggle for existence.") Then, from natural selection we move to the division of labor and the generation of different forms. As Newton explains Copernicus – the heliocentric universe – Darwin explains Genesis – the tree of life! This is what being a *vera causa* is all about.

Consilience

Moving to the third part of the *Origin*, we begin (Chapters V and VI) with some clean-up issues, such as the nature of variation and some difficulties with and misunderstandings of the nature of selection. Darwin reaffirmed that homology is a result of selection but not designed by selection. "It is generally acknowledged that all organic beings have been formed on two great laws – Unity of Type, and the Conditions of Existence." He explains that by "unity of type" one is talking of the shared structure between organisms of different species. This shared structure has nothing to do with how organisms live their lives. "On my theory, unity of type is explained by unity of descent" (206). Elaborating: "Nothing can be more hopeless than to attempt to explain this similarity of pattern in members of the same class, by utility or by the doctrine of final causes. The hopelessness of the attempt has been expressly admitted by Owen in his most interesting work on the 'Nature of Limbs'" (435) (Figure 3.3). Organisms start off as one group, then they separate and

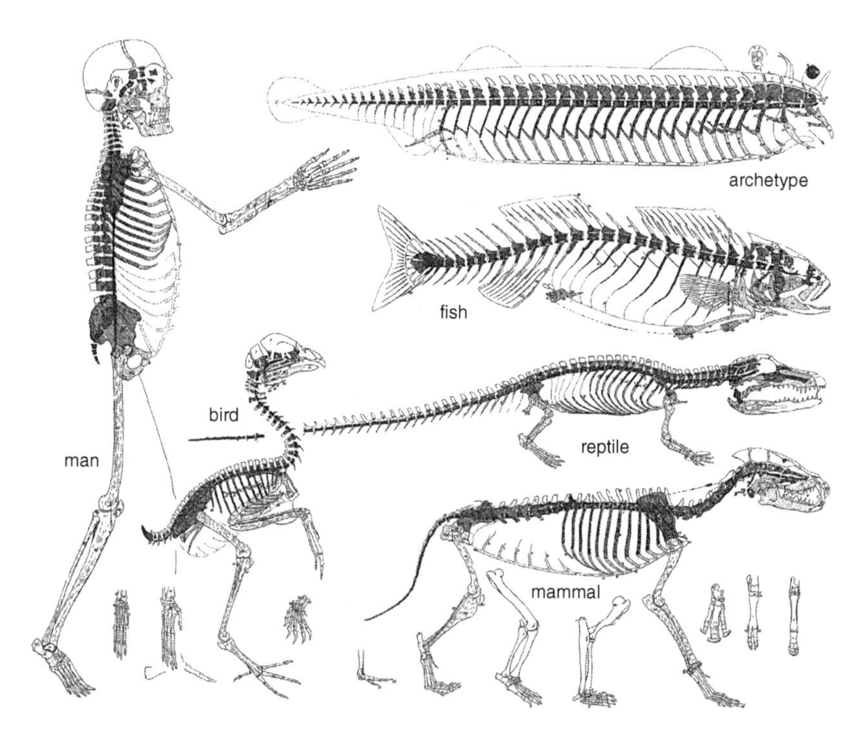

Figure 3.3 Richard Owen's *On the Nature of Limbs* (1849), showing the vertebrate archetype, and thus the "homologies" between different organisms. (For easier reading and interpretation, the diagram has been simplified.)

evolve independently, but continue to show evidence of their shared history. One implication of this is that Darwin never claimed that every aspect of organisms is adaptive. Natural selection may be all important; but it may, in the pursuit of adaptive efficiency, produce nonadaptive features as a side effect. As long as they don't unduly disturb the adaptive features, natural selection does not stop their persisting. Or stand in the way of their being the crucial factors in determining relationships.

For the rest of the *Origin* (Chapters VII through XIV) we get a perfect example of a Whewellian consilience of inductions, with the *vera causa* of natural selection tying all together. We start with "Instinct" (VII). In his response to Wallace, Darwin speaks of a "social community." In looking first at social behavior, he explains what he means. In breeding, if they want to get the desirable features of the slaughtered animal, farmers know that they can return to the family. So sterile workers in nests of ants come about for the same reason. They have advantages for the interrelated group, so they are perpetuated. The group is not just individuals, but a family. First, the problem: "I can see no real difficulty in any character having become correlated with the sterile condition of certain members of insect-communities: the difficulty lies in understanding how such correlated modifications of structure could have been slowly accumulated by natural selection" (237). Then, the solution: "This difficulty, though appearing insuperable, is lessened, or, as I believe, disappears, when it is remembered that selection may be applied to the family, as well as to the individual, and may thus gain the desired end." What is the evidence for this claim? Breeders of plants and animals know that by turning back to the original stock, you can keep a line going indefinitely. Even though the features you want are always in organisms that never breed, existing only for human consumption (237–38).

Finally, the application:

> Thus I believe it has been with social insects: a slight modification of structure, or instinct, correlated with the sterile condition of certain members of the community, has been advantageous to the community: consequently the fertile males and females of the same community flourished, and transmitted to their fertile offspring a tendency to produce sterile members having the same modification. (238)

Moving on briskly through the other areas, next comes "paleontology" – the fossil record (Figure 3.4). The gaps in the record are apparently a function of nondeposition rather than nonexistence. More positively, the overall record is explained in terms of descent with modification. How else is one going to explain what Lyell suspiciously leaves silent, that fossil forms are not randomly

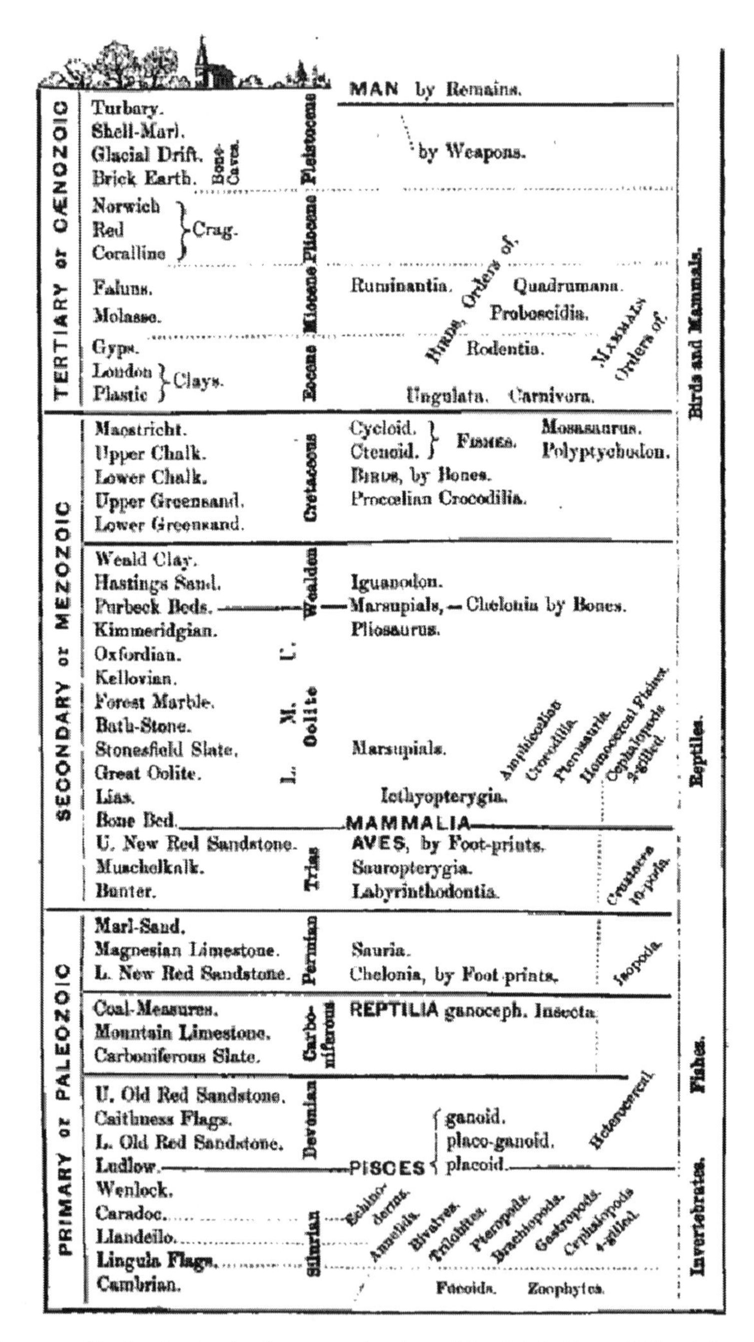

Figure 3.4 The fossil record as known at the time of the *Origin*. From Richard Owen's *Paleontology* (1860).

distributed but that new forms are frequently – usually – similar to forms only a little further down the record, and thus little more recent than the older forms? Darwin also takes up the common finding that the fossils of long-extinct organisms often seem to link organisms very different today. In language already encountered, ancient forms show the archetype that today is put to very different uses – wings, fins, legs, arms – by today's organisms. "It is a common belief that the more ancient a form is, by so much the more it tends to connect by some of its characters groups now widely separated from each other" (330). This is hard to prove generally: "Yet if we compare the older Reptiles and Batrachians, the older Fish, the older Cephalopods, and the eocene Mammals, with the more recent members of the same classes, we must admit that there is some truth in the remark" (330–31). Evolution offers the explanation.

Vestiges, like Erasmus Darwin (and Lamarck for that matter), made much of the progressive nature of the fossil record. As noted earlier, Charles Darwin was always far cagier on this issue. He was a student of Lyell, whose geological picture was built on a steady-state world. Lyell in turn followed the system of the eighteenth-century Scottish geologist James Hutton, "epitomized by its conclusion 'we find no vestige of a beginning – no prospect of an end'" (Rudwick 1970, 8–9). From the start, Darwin saw that natural selection is relativistic. When food is abundant, being big and able to ward off predators is of value. When food is scarce, being small and able to avoid danger is of value. That said, Darwin recognized a flavor of progress in the record. "There has been much discussion whether recent forms are more highly developed than ancient" (336–37). Darwin admitted that he was not sure about this, because no one really knows what is meant by "higher" and "lower." Yet "in one particular sense the more recent forms must, on my theory, be higher than the more ancient; for each new species is formed by having had some advantage in the struggle for life over other and preceding forms" (337). He added, somewhat gloomily, "I can see no way of testing this sort of progress." Picking up on our discussion of values in Chapter 1, we have seen Darwin offering an analysis of final cause that does not appeal to values – no Mind behind it all that "would direct everything and arrange each thing in the way that was best" (Cooper 1997: *Phaedo*, 97 c–d). Nor equally a force directing organisms towards a Perfect Being. Now, in this discussion, he is pulling back from an inbuilt upwards motion, value-impregnated in aiming towards the highest organisms, humankind. Darwin's theory is a mechanist's view of the living world. No built-in values.

Progressive or not, many think that the truth of evolution lies with the fossil record. Not for nothing did the American Creationist Duane T. Gish call his anti-evolution tract *Evolution: The Fossils Say No!* (1973). To the contrary – as one might expect from comments in his *Journal of Researches* – Darwin, and professional evolutionists following in his path, regarded "Geographical distribution" (Chapters XI and XII) as a very strong point, if not the strongest. If there is no evolution, then why did God so love the Galapagos as to furnish the islands with their own species of bird and tortoise? (Refer here back to Figure 2.4.) And why are the birds of the Galapagos like the birds of South America rather than the birds of Africa? And why does the converse also hold true of the birds of the Cape Verde islands of the Atlantic with respect to Africa and South America? "There is nothing in the conditions of life, in the geological nature of the islands, in their height or climate, or in the proportions in which the several classes are associated together, which resembles closely the conditions of the South American coast: in fact there is a considerable dissimilarity in all these respects" (398). Adding:

> On the other hand, there is a considerable degree of resemblance in the volcanic nature of the soil, in climate, height, and size of the islands, between the Galapagos and Cape de Verde Archipelagos: but what an entire and absolute difference in their inhabitants! The inhabitants of the Cape de Verde Islands are related to those of Africa, like those of the Galapagos to America.

How else can one explain this except through descent with modification?

And so we move on towards the end of the consilience, covering "Classification," "Morphology," "Embryology," and "Rudimentary Organs" (XIII). Classification or Systematics reflects the history of life: "the natural system is founded on descent with modification" (420). Morphology, the Unity of Type. "The explanation is manifest on the theory of the natural selection of successive slight modifications,—each modification being profitable in some way to the modified form, but often affecting by correlation of growth other parts of the organisation. In changes of this nature, there will be little or no tendency to modify the original pattern, or to transpose parts" (435). Going out of order, Rudimentary Organs are easily explained as features that were of little value and so evolution often left them behind, not bothering to eliminate them. Far more interesting is Embryology, a particular favorite of Darwin. He saw the nature and development of embryos to be a function of the ages at which natural selection becomes active. If the embryos and adults of two different species are different all the way, then one presumes that the forces of selection are active all through the organisms' development. But if, as frequently happens, there is much similarity between the embryos of organisms very different as adults, then one suspects that selection only came into play when the organisms were born. Up to then, there is no reason to think that selection is tearing them apart. That only happens when they have left the security of the womb or its equivalent (Figure 3.5).

Chap. I. EMBRYONIC DEVELOPMENT.

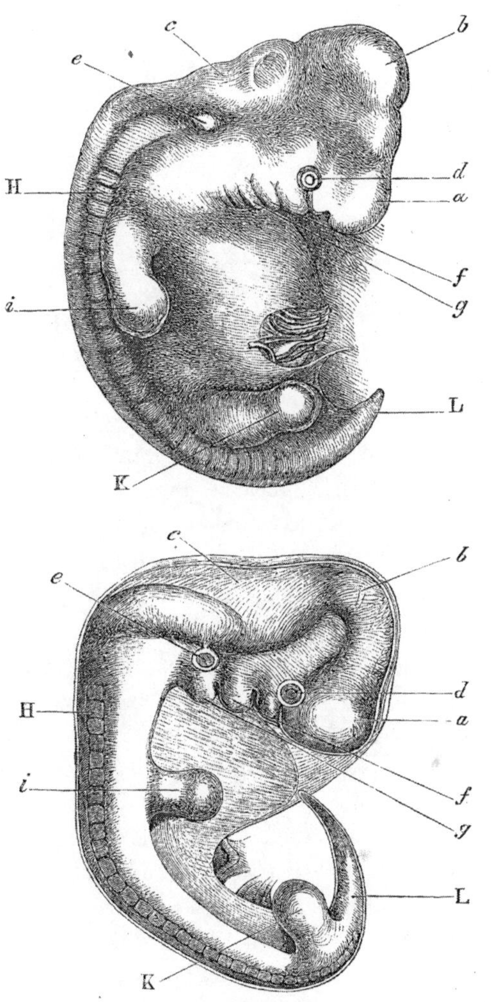

Fig. 1. Upper figure human embryo, from Ecker. Lower figure that of a dog, from Bischoff.

a. Fore-brain, cerebral hemispheres, &c.
b. Mid-brain, corpora quadrigemina.
c. Hind-brain, cerebellum, medulla oblongata.
d. Eye.
e. Ear.
f. First visceral arch.
g. Second visceral arch.
H. Vertebral columns and muscles in process of development.
i. Anterior } extremities.
K. Posterior } extremities.
L. Tail or os coccyx.

Figure 3.5 Embryos of human (above) and dog (below), showing the similarity of embryos of organisms very different as adults. From *The Descent of Man* (1871).

Darwin invoked the analogy with human-caused change. "Some authors who have written on Dogs, maintain that the greyhound and bulldog, though appearing so different, are really varieties most closely allied, and have probably descended from the same wild stock..." (444–45). Darwin was not about to take this conclusion on trust:

> I was curious to see how far their puppies differed from each other: I was told by breeders that they differed just as much as their parents, and this, judging by the eye, seemed almost to be the case; but on actually measuring the old dogs and their six-days old puppies, I found that the puppies had not nearly acquired their full amount of proportional difference. (445)

Just what one would expect if Darwin's theory about embryological development is correct. Breeders care about the form of the adult dogs, not the puppies.

Drawing the *Origin* to a close with "Recapitulation and Conclusion" (XIV), Darwin touched very briefly on the evolution of a species hitherto-undiscussed – *Homo sapiens*. We have seen that, from the first, Darwin was convinced that we are part of the whole. But he did not want this to dominate the discussion, at least not at this stage. It was obviously going to be a matter of great controversy – it was, with Darwin's theory known as the "monkey theory" (Figure 3.6). Time to make the point and move on. "In the distant future I see open fields for far more important researches. Psychology will be based on a new foundation, that of the necessary acquirement of each power and capacity by gradation. Light will be thrown on the origin of man and his history" (488). Even more important than mentioning our species was to affirm the compatibility of his theory with religious belief. As we have seen, *Vestiges* offers a kind of natural theological argument that a lawbound origin of species makes for a more glorious god than one who has to create, over and over again, by miracle. Darwin has a similar argument, but since it appears in the *Sketch*, scribbled two years before *Vestiges*, it is not a copy. Remaining unchanged through the six editions, Darwin wrote: "When I view all beings not as special creations, but as the lineal descendants of some few beings which lived long before the first bed of the Silurian system was deposited, they seem to me to become ennobled" (488–89). Although he himself was moving to agnosticism, in the *Origin* Darwin kept open the possibility of a Platonic Creator. One who was "hands on."

The *Origin* drew to a close. The conclusion, echoing both the *Sketch* and the *Essay*: "There is grandeur in this view of life, with its several powers, having been originally breathed into a few forms or into one; and that, whilst this planet has gone cycling on according to the fixed law of gravity, from so

Figure 3.6 Darwin as a monkey, cartoon in *The Hornet*, March 22 1871.

simple a beginning endless forms most beautiful and most wonderful have been, and are being, evolved" (490).

Although Darwin was trying to avoid values in his science, it was not always easy. He had the values. The problem was that of keeping them out of his science. With echoes of the Copernican Revolution – "this planet has gone cycling on according to the fixed law of gravity" – his *Newtonian* science.

Paradigm Change?

So much for the theory of the *Origin*, the heart of the "Darwinian Revolution." Which at once raises the fact that no one could write a book on a scientific revolution – certainly not on the Darwinian Revolution – without having in mind Thomas Kuhn's celebrated book *The Structure of Scientific Revolutions* (1962). He argues that such changes, between what he refers to as "paradigms" – akin to theories but somewhat broader, incorporating sociological factors such as training – are not entirely reason-driven, as would be argued by someone such as Karl Popper (1959). Paradigms demand a kind of commitment, like a religious position or a political calling. That is reflected in the change from one paradigm to another. Such change requires something akin to a leap of faith. Apart from anything else, paradigms are "incommensurable." They belong to different worlds. This, as with religion and politics, explains why often there are somewhat violent reactions when paradigm changes occur. Those in the older paradigm can take extreme umbrage at the claims and enthusiasms of those in the newer paradigm.

This book is not being written as a test case for Kuhn's thinking, but that does not stop us from using Kuhn's thinking to see if it can help us understand the theory presented in this chapter. A pressing imperative, because some of the things that Kuhn says about paradigm change do apply somewhat uncannily to the Darwinian Revolution. We start with the claim, stressed by Kuhn, that paradigm shifts don't usually occur unless the old paradigm is showing its age, with uncomfortable facts unexplained. In the Darwinian case, the old paradigm made much of the Designing nature of God and how this is shown throughout our understanding of the external world, especially the external world of animals and plants. And yet all agreed that homology was an unsolved issue. There just seems to be no reason why so many vertebrates are guided by shared templates – shared templates that are put to different functional uses: the forelimb of the horse, the arm of humans, the wings of birds, the fins of whales. Richard Owen's little book *On the Origin of Limbs* (1849) made all this dreadfully clear, an uncomfortable conclusion, for there are hints that already Owen was moving towards evolution. The best that someone such as Whewell could say was that it was no big problem: "the modern doctrine of Unity of Plan in different kinds of animals does not at all necessarily contradict the Doctrine of Final Causes: ... Morphology is not necessarily inconsistent with Teleology" (Whewell 1845, 9). Which may be true but is hardly a reasoned solution. Why isn't morphology necessarily inconsistent with teleology?

This was Whewell failing to deal with an already existing problem. In the decade before the *Origin*, Whewell helped to increase the problem. In the early 1850s, he got into controversy about the possibility of extraterrestrial intelligent beings. His book, *On the Plurality of Worlds*, was published anonymously, but everyone knew the identity of the author (Whewell [1853] 2001). Are there such beings? If there are, does this mean that Jesus has to go around the universe, atoning for the sins of such beings? Every Friday, a crucifixion somewhere in the universe? Or are we unique, in which case why did God create all the other planets? Whewell's implausible suggestion was that the other planets were created to give us an opportunity to be God-like. Man's mind is in essential respects like God's Mind, and part of our task on Earth might be to bring ourselves closer to God by tracing His laws as manifested by the endless motions of the heavenly bodies. If this be true, "we cannot have any ground to think that the scheme of creation is too narrow; or that it needs, in order to give it sufficient dignity and value, and a worthy object in our eyes, that other worlds should be stocked with races of creatures…" (309).

Expectedly, this solution did not bring Whewell unalloyed praise. The most enterprising critic was Sir David Brewster, Scottish physicist and biographer of Newton. In his gloriously labeled *More Worlds than One: The Creed of the Philosopher, and the Hope of the Christian* (1854), at the revealed theological level, Brewster was keen to show that the Bible strongly supports extraterrestrials. Apparently, Psalm 7 pointed that way: "When I consider thy heavens, the work of thy fingers, The moon and the stars, which thou hast ordained; What is man, that thou art mindful of him?" There was much more where that came from. At the natural theological level, laboriously, the whole universe is shown to be inhabited. Even the sun! "The probability of the sun being inhabited is doubtless greatly increased by the simple consideration of its enormous size." Continuing that "it is difficult to believe that a globe of such magnificence, 882,000 miles in diameter, upwards of one hundred and eleven times the size of our earth, and 1,384,472 times its bulk, should occupy so distinguished a place without intelligent beings to study and admire the grand arrangements which exist around them" (Brewster 1854, 102). It is hard not to conclude that, whatever knots Whewell was tying himself in, they are small compared with the snares that Brewster was preparing for himself. One doubts that even the imagination of H. G. Wells would have succeeded in making plausible supergeniuses live happily in the flaming center of our universe. Whewell's problems with normal science were nothing compared with those of Sir David.

So much to the positive side to Kuhn's analysis. Turn now to the question of paradigm change, supposedly from one incommensurable picture of the world to another. At one level, the Darwinian Revolution seems to back this, most directly in the tensions that Darwin's theory raised for the old guard. Herschel referred to natural selection as the law of "higgledy-piggledy." Whewell, now Master of Trinity, would not allow, in the college library, a copy of so dangerous a book, so great a threat to the mental wellbeing of naïve undergraduates. Sedgwick, expectedly, went over the top:

> If I did not think you a good tempered & truth loving man I should not tell you that, (spite of the great knowledge; store of facts; capital views of the corelations of the various parts of organic nature; admirable hints about the diffusions, thro' wide regions, of nearly related organic beings; &c &c) I have read your book with more pain than pleasure. Parts of it I admired greatly;

> parts I laughed at till my sides were almost sore; other parts I read with absolute sorrow; because I think them utterly false & grievously mischievous. (Darwin 1985-, 7: Letter from Sedgwick to CD, November 24, 1859)

However, when we start to dig into the incommensurability issue, things start to fall apart. Just about every element in Darwin's theory seems to have been appropriated from the culture in which he lived. Darwin grew up in Shrewsbury, the very heart of agricultural England. *Breeding* was mentioned in the Bible, as Darwin noted in the *Essay*: "In the earliest chapters of the Bible there are rules given for influencing the colours of breeds, and black and white sheep are spoken of as separated" (Darwin 1909, 67). By the time of the *Origin*, Darwin realized that there was no need to invoke the authority of Scripture. In the nineteenth century, animal husbandry was a huge enterprise, as, thanks to the Industrial Revolution, farmers had to speak to the need for greatly increased amounts of food (for the expanding urban population), with much decreased available labor to produce it (thanks to the exodus from rural areas to urban areas). Selective breeding was the key – fatter pigs, beefier cattle, and more. Darwin soaked this up, as we have seen from his reading of Sebright.

Variation was made apparent to Darwin while he worked on barnacles, work which was almost ostentatiously pre-Darwinian, with its troubles from adaptation. The *struggle for existence* came straight from Malthus. Darwin used it for his own purposes. Malthus put the struggle in a natural theological context, as God's way to get us to work hard for a living. Something is needed to spur philosophy grad students to get out of bed in the morning. However, whatever its intent, Darwin's appropriation of the concept was nevertheless pure Malthus. Moving on, *adaptation* was design, design, design. It was the central focus of the textbook *Natural Theology* (1802, collected edition 1819), by Archdeacon William Paley. A work Darwin studied intensively while at Cambridge, the central argument is that the eye is like a telescope, and as telescopes have designers, so likewise the eye has a designer – or Designer – the Great Optician in the Sky. In the words of Whewell (1840): "There is one idea which the researches of the physiologist and the anatomist so constantly force upon him, that he cannot help assuming it as one of the guides of his speculations; I mean, the idea of a *purpose*, or, as it is called in Aristotelian phrase, a *final cause* came, in the arrangements of the animal frame" (my emphasis).

"There are few more impressive sights in the world than a Scotsman on the make" (James Barrie 1908). The *division of labor* is pure Adam Smith (1776), a combination of Scottish economics and Presbyterian Christianity. It sounds like John Knox to say that, by everyone pursuing their own interests, the world generally is going to be made better. And then there is the *tree of life*, the Darwinian counterpoint to the effects of Newtonian gravitation: Kepler's laws of heavenly motion and Galileo's laws of terrestrial motion. At the beginning of the Bible: "And out of the ground made the LORD God to grow every tree that is pleasant to the sight, and good for food; the tree of life also in the midst of the garden, and the tree of knowledge of good and evil" (Genesis 2:9). And, at the end of the Bible: "He that hath an ear, let him hear what the Spirit saith unto the churches; To him that overcometh will I give to eat of the tree of life, which is in the midst of the paradise of God" (Revelation 2:7).

As has been stressed, Darwin was a very comfortable member of the upper-middle classes, living in a hugely successful society, to which the Darwin/Wedgwood family contributed and from which the family benefited. It would need explanation were he not to draw on his culture in devising his theory. Especially one such as he, with ongoing sickness, and hence confined to his house, his reading, his correspondence. So here the Darwinian Revolution was very much not a Kuhnian Revolution. But before we turn away, think again about Kuhn's strong point, the emotional rejection of the *Origin* by intelligent and informed people such as Sedgwick. There was something there that upset them. I suggest that a far better explanation than incommensurability is the metaphor of a *kaleidoscope*. It takes elements, shakes them up, and produces an entirely new picture. This is the way to consider the theory of the *Origin*. It was new. It was revolutionary. Naturally, it upset Sedgwick. Apart from anything else, there was no direct evidence of natural selection in motion. It was just not the work of a rebel. Darwin was the quintessential Englishman, and his theory is the theory that would be produced by a quintessential Englishman. Meaning it was a theory from a man who was immersed in English culture.

4

Evolution in the Nineteenth Century

Oxford

On June 30, 1859, seven months after the publication of the *Origin of Species*, at the annual meeting of the British Association for the Advancement of Science, in Oxford, the High-Church bishop of that city, Samuel Wilberforce – known as "Soapy Sam," son of the fighter against slavery, William Wilberforce – faced off against Thomas Henry Huxley, professor of natural history at the Royal School of Mines, and self-styled "Darwin's bulldog" (Figure 4.1). Supposedly, the bishop asked the professor whether he was descended from apes on his grandfather's side or his grandmother's. The professor replied that he would sooner be descended from a humble ape than from a bishop of the Church of England. Everyone was having a wonderful time, although sadly the story is undoubtedly embellished (Lucas 1979). But this is one of those times when fiction tells you more than fact. Darwin had changed the rules of the game. No longer was the idea of evolution something almost certainly false and quite certainly rather vulgar. It was now a hypothesis that all could consider and debate. Causes apart, Darwin's splendid review of the evidence – instinct, paleontology, biogeography, morphology, systematics, embryology – meant that evolution had arrived. Whether or not you accepted the hypothesis, evolution was now part of the scene.

Organicism Redivivus

By the mid nineteenth century, as the Oxford confrontation clearly shows, mechanism was proving its worth again and again. In Darwin's *Origin*, he set out to give the biological equivalent of Newtonian mechanics, the final stage of the effort to show that the world could be explained by scientific theories guided by the machine root metaphor. He accepted teleology. Then,

Figure 4.1 Wilberforce and Huxley.

he offered an account of teleology that fell under this metaphor. And above all, he strove to keep values out of his science, most particularly in repudiating claims about progress and humans at the top. From the beginning of his thinking about evolution, as soon as he discovered natural selection, Darwin was arguing that it gives no guarantee of progress. What else would one expect from someone so hugely within Lyell's uniformitarian orbit? From an early notebook: "The enormous number of animals in the world depends of their varied structure & complexity. – hence as the forms became complicated, they opened fresh means of adding to their complexity. – but yet there is no necessary tendency in the simple animals to become complicated although all perhaps will have done so from the new relations caused by the advancing complexity of other" (Notebook E97. Written in January 1839). On the flyleaf of his copy of *Vestiges,* he cautioned himself never to use the terms "higher" and "lower" (Di Gregorio and Gill 1990).

Darwin kept on worrying about this issue. We have seen that, in the first edition of the *Origin*, 1859, he does allow a kind of progressive odor to the fossil record, but it is hardly an enthusiastic endorsement. In the third edition of the *Origin*, 1861, just two years after the first edition, he added several new paragraphs on the topic. He basically repeated the sentiment in his notebooks about organization leading to highness. "If we look at the differentiation and specialisation of the several organs of each being when adult (and this will include the advancement of the brain for intellectual purposes) as the best standard of highness of organisation, natural selection clearly leads towards highness…" (Darwin 1861, 134). But then, later – in this same edition – he qualified what he had said to be virtually vacuous: "To attempt to compare in the scale of highness members of distinct types seems hopeless: who will decide whether a cuttlefish be higher than a bee—that insect which the great Von Baer believed to be 'in fact more highly organised than a fish, although upon another type'?" Unhelpfully, he added: "In the complex struggle for life it is quite credible that crustaceans, for instance, not very high in their own class, might beat the cephalopods or highest molluscs; and such crustaceans, though not highly developed, would stand very high in the scale of invertebrate animals if judged by the most decisive of all trials—the law of battle" (365). Darwin kept emphasizing the underlying sentiment even after the *Descent* was published. To the American evolutionist Alphaeus Hyatt, he wrote: "After long reflection I cannot avoid the conviction that no innate tendency to progressive development exists, as is now held by so many able naturalists, & perhaps by yourself" (Darwin 1985: Letter, December 4, 1872).

Yet there were always those who regretted the demise of the organic root metaphor. Somehow there was a feeling that something of value had been lost. Something of spiritual value, without necessarily being overtly Christian. Particularly in Germany, there was a resurgence of organicism at the end of the eighteenth century, promoted by the "Romantics" (Richards 2002; Rupik 2024). The poet Goethe was a prominent figure. He dabbled in the life sciences, known particularly for his vision of the plant as an *Urpflanz*, where there is a unity imposed by all parts being manifestations of the same underlying pattern – something akin to a Platonic Form (Goethe 1790) (Figure 4.2). Goethe is better known, but it was the philosopher Friedrich Schelling who really ran with this philosophy. As a teenager, he wrote a sixty-page essay on the *Timaeus*, and that early experience colored all his later thinking: "Even in mere organized matter there is life, but a life of a more restricted kind" (Schelling 1797, 35). The world's nature is such that it produces itself, its development comes from within, as an unfurling organism is produced by forces within rather than without. The acorn naturally grows up into the oak. From the simple to the complex, from the undifferentiated to the highly differentiated. Mind and body are one: "Nature should be Mind made visible, Mind the invisible nature. Here then, in the absolute identity of Mind in us and Nature outside us, the problem

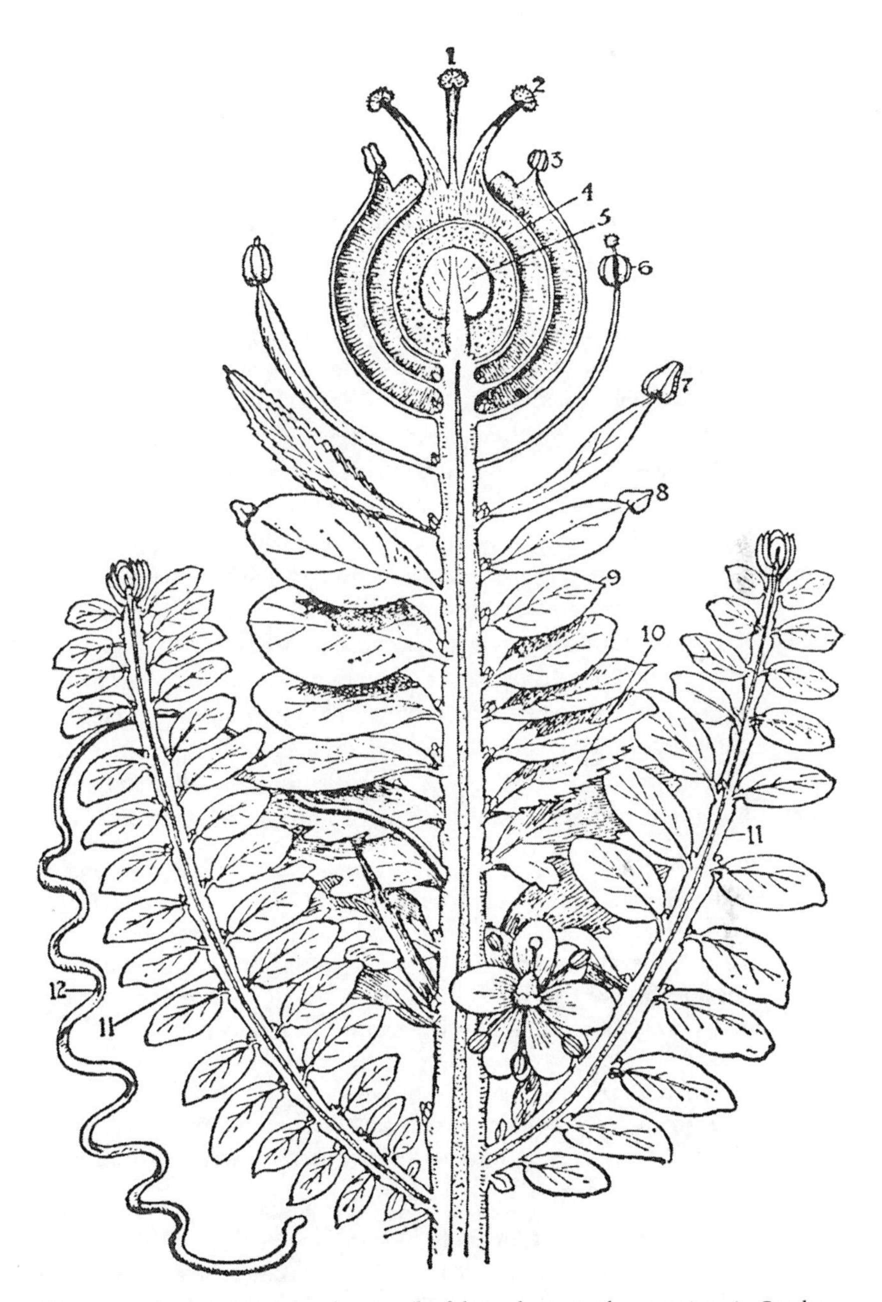

Figure 4.2 Goethe's *Urpflanz*, showing (leaf) homologies in the organism, in Goethe, *On the Metamorphosis of Plants* (1790).

of the possibility of a Nature external to us must be resolved" (42). This was a form of pantheism. (Pantheism, as in Spinoza, identifies mind with body. As a Platonist, Schelling would have seen the Form of the Good as separate from body. It embraces body with its spirit. This is known as "panentheism.")

In a way not true of mechanism, as noted earlier with the talk of acorns to oaks, evolution – self-driven, progressive evolution ending with humans – practically comes with organicism. One who responded positively to this kind of thinking was the German evolutionist Ernst Haeckel (1834–1919). He claimed – he genuinely thought – he was a faithful disciple of Darwin. But if one looks carefully at his writing and theorizing, he sounds much more Romantic – organismic – than Darwinian – mechanistic (Richards 2008). This came naturally to one who cut his scientific teeth on embryological studies – the area of biology focusing on the development, irrespective of outside forces, of the fertilized egg to the full-grown adult. This hints – more than hints – that biological development, change of any kind, is going to originate from within, as it were, rather than from without, which latter is precisely the way that the force of natural selection works. Little wonder that Haeckel championed the "biogenetic law": "ontogeny recapitulates phylogeny" (Haeckel 1866). As the individual organism develops, it is precisely mimicking the way that the group develops. It generates its own value rather than needing us to read in such value. Confirming Haeckel's debts to Romanticism, in looking at the many phylogenies that he drew – he was a talented illustrator – we see progress, especially progress to human beings. Again, value is found, not imputed (Figure 4.3). Complementing this, he wrote:

> Our Theory of Development explains the origin of man and the course of his historical development in the only natural manner. We see in his gradually ascensive development out of the lower vertebrata, the greatest triumph of humanity over the whole of the rest of Nature. We are proud of having so immensely outstripped our lower animal ancestors, and derive from it the consoling assurance that in future also, mankind, as a whole, will follow the glorious career of progressive development, and attain a still higher degree of mental perfection. When viewed in this light, the Theory of Descent as applied to man opens up the most encouraging prospects for the future, and frees us from all those anxious fears which have been the scarecrows of our opponents. (Haeckel 1868, 945)

Stress again, Haeckel's vision of evolution was value-laden – value-laden, with humans at the top of the tree. Christian critics could be answered in a crushing manner. The evolutionist privileged humans no less than the author of Genesis and gave reasons why we humans are privileged. Most simply did not want to go the way of Darwin, explicitly eschewing progress and thereby removing one of the strongest planks supporting the Temple of Evolution.

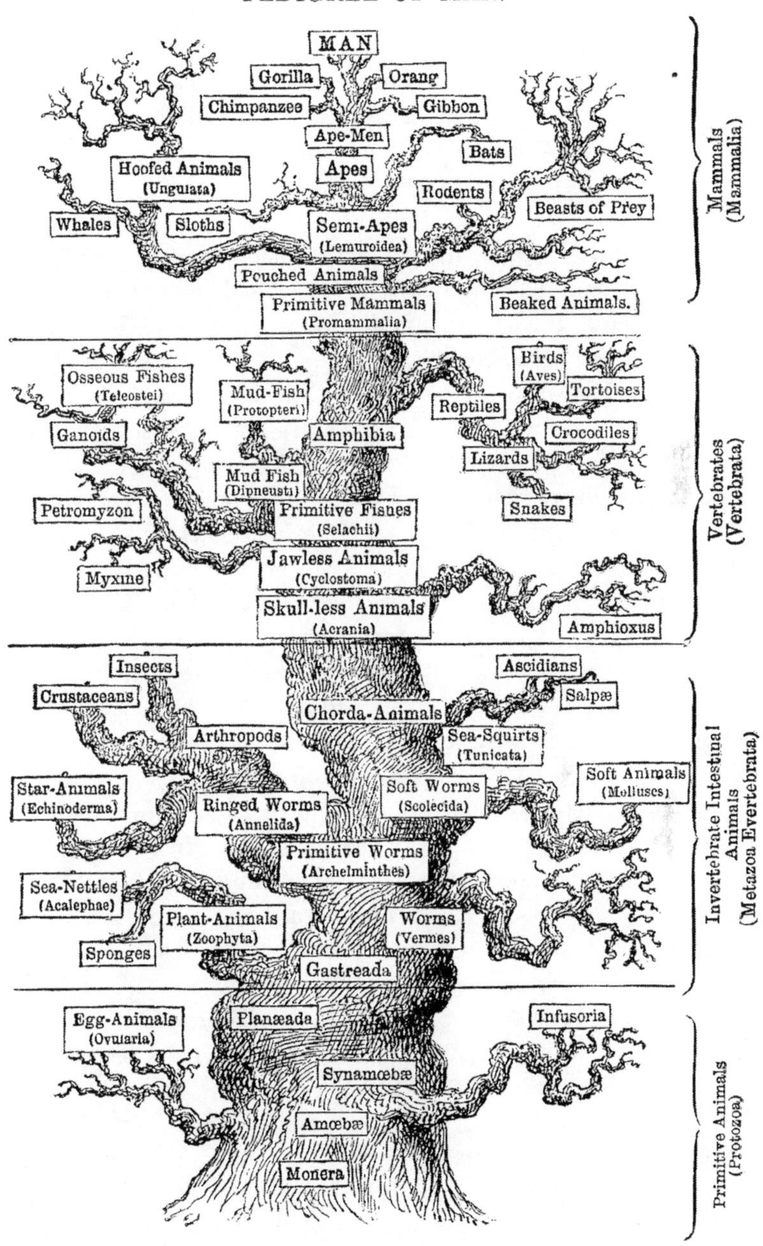

Figure 4.3 The Tree of Life as drawn by Ernst Haeckel. Note that it is progressive, with humans at the top. Haeckel, *Anthropogenie oder Entwickelungsgeschichte des Menschen* (1874).

Herbert Spencer

Britain had its own version of Romantic-influenced evolution (Peel 1971; Wiltshire 1978; Francis 2007). The maverick student of just about everything, Herbert Spencer, acknowledged explicitly his debt to Schelling, via the plagiaristic writings of Coleridge (Duncan 1908). Born in the British Midlands, ten years younger than Darwin, nonconformist (Protestant, non-Anglican – his father moved from Methodism to Quakerism), middle-class although at a much lower level than Darwin, Spencer had a somewhat eclectic upbringing and education, finally learning some mathematics and Latin from an Anglican priest uncle. Unable – unwilling – to settle into regular employment (he worked a little surveying in the railway boom of the late 1830s), he became a writer and editor, mainly producing material for radical provincial magazines. Working, from 1848 to 1853, as a subeditor of the free-trade journal *The Economist*, Spencer became known to a group of (what we would call) left-wing intellectuals, including John Stuart Mill, George Lewes, and Mary Ann Evans (later the novelist George Eliot), with whom apparently Spencer had a brief romance. (Later Lewes and Evans became partners, unable to marry because Lewes already had a wife.)

Having picked up and imbibed bibs and bobs of evolutionary thinking, Spencer announced – bellowed forth – his beliefs, almost ten years before the *Origin* was published (although ten years or more after Darwin had become an evolutionist and discovered natural selection). Impervious to all criticism, Spencer had little time for opponents, dismissing them as having no real alternative view of life: "Those who cavalierly reject the Theory of Evolution as not being adequately supported by facts, seem to forget that their own theory is supported by no facts at all. Like the majority of men who are born to a given belief, they demand the most rigorous proof of any adverse belief, but assume that their own needs none" (Spencer 1852b, 377). Although he hit (independently) on the idea of natural selection, Spencer was truly committed to Lamarckism, which he wove into his own idiosyncratic theory. For all that Darwin always accepted Lamarckism as a secondary mechanism, Spencer's evolutionism was very different from that of Darwin. For the great naturalist, evolution is something to be proven by empirical evidence. For Spencer, it comes with the philosophy. Darwin saw Malthusian pressures as leading to a situation where those with certain variants, helpful to survival and reproduction, actually succeeded better than their rivals. Spencer saw the Malthusian pressures as putting demands on organisms to develop better characteristics through Lamarckian processes (1952a). As organisms adapt to their circumstances, reproductive necessities decline, as the chances of successful offspring increase. Spencer himself was so far advanced that he had no offspring at all. In short, for Darwin, evolution was a population issue, with change coming from external forces sorting through the groups (not group selection, for the successes of the successful came from their own efforts), whereas Spencer (influenced by Schelling) saw the changes coming from within individual organisms: organicist rather than mechanist. Expectedly, there was the embrace of value within nature, discovered rather than imposed. Spencer was indifferent to the philosophy underlying the message of the *Origin*. Happily, he stressed in no uncertain terms the progressive nature of just about everything:

> Now we propose in the first place to show, that this law of organic progress is the law of all progress. Whether it be in the development of the Earth, in the development of Life upon its surface, in the development of Society, of Government, of Manufactures, of Commerce, of Language, Literature, Science, Art, this same evolution of the simple into the complex, through successive differentiations, holds throughout. From the earliest traceable cosmical changes down to the latest results of civilization, we shall find that the transformation of the homogeneous into the heterogeneous is that in which Progress essentially consists. (Spencer 1857, 2–3)

For Spencer, there were no exceptions to this law. With respect to other animals, humans were more complex or heterogeneous; with respect to (what he would have thought of as) savages, Europeans more complex or heterogeneous; and with respect to the tongues of other peoples, the English language more complex or heterogeneous.

Spencer combined this with an emphasis on the wholeness of things, rather than entities best broken down for greater understanding: "Societies slowly augment in mass; they progress in complexity of structure; at the same time their parts become more mutually dependent; their living units are removed and replaced without destroying their integrity; and the extents to which they display these peculiarities are proportionate to their vital activities." Developing the analogy:

> These are traits that societies have in common with organic bodies. And these traits in which they agree with organic bodies and disagree with all other things, entirely subordinate the minor distinctions: such distinctions being scarcely greater than those which separate one half of the organic kingdom from the other. The *principles* of organization are the same, and the differences are simply differences of application. (Spencer 1876, 1)

Spencer combined all of this in an overall picture he called "dynamic equilibrium": "that state of organic moving equilibrium which we saw arises in the course of Evolution, and tends ever to become more complete" (Spencer 1864, 1, 93). Influenced by the second law of thermodynamics, he argued that groups of organisms, individuals in their own right, will be in a state of equilibrium. This is always unstable and so breaks down. In coming together again and reachieving equilibrium, we move from the homogeneous to the heterogeneous. Progress!

Thomas Henry Huxley

Although, thanks to the family fortune, Darwin never had to work for a living, no one denied his status as a professional biologist. The work on barnacles secured this status; although really, by the end of the 1830s, with his increasing number of publications, his Royal Society membership, and his place as a member of the board of the Geological Society, Darwin was fully accepted as professional. Not so with Spencer, who (to be honest) never strove for such a status. He was ever a general man of letters. In the world of science, increasingly this told against him – in Britain at least. Spencer's good friend, the anatomist (who was developing as a paleontologist) Thomas Henry Huxley, at the beginning of the 1850s newly returned from four years as surgeon on a British warship somewhat copying the *Beagle* as it explored the East Indies and the coast of Australia, was, from his arrival back in Britain, striving nonstop to establish science as an enterprise of secular activity – research and teaching – that could support its practitioners (Desmond 1998; Ruse 1996). In 1854, he left the navy and became (as mentioned) the Professor of Natural History at the Royal School of Mines, a post he kept for over thirty years. By the 1860s, Huxley and like thinkers (which included people such as Darwin's friend Joseph Dalton Hooker, now based at Kew Gardens) were starting to have considerable success. The Royal School of Mines moved to South Kensington, and Huxley was both a teacher and an administrator at what today has evolved into the Imperial College of Science.

People like Spencer got rather pushed aside. They did not fit into the pattern of the new kind of scientist – professional and supported because of their learning, talents, and activities. One should stress that this did not happen because people doubted Spencer's evolutionism. Huxley is a good contrast. He was a hugely enthusiastic evolutionist, and, as soon as the *Origin* was published, he wrote a glowing review in the establishment newspaper, *The Times.* Appearing on December 26, in England a national holiday, Boxing Day, the respectable middle classes, eating the leftovers from the day before, would have learnt that evolution was no longer the property of the radicals and the gullible:

> Mr. Darwin abhors mere speculation as nature abhors a vacuum. He is as greedy of cases and precedents as any constitutional lawyer, and all the principles he lays down are capable of being brought to the test of observation and experiment. The path he bids us follow professes to be, not a mere airy track, fabricated of ideal cobwebs, but a solid and broad bridge of facts. (Huxley 1893b, 20–21)

However, for Huxley, belief in evolution was not an end in itself – as it was for both Spencer and Darwin – but a tool to promote a naturalist vision of science, moving on from the science (in England) of the 1830s, controlled by the Anglicans from the ancient universities. Huxley was no atheist – it was he who invented the term "agnosticism," as a position between belief and its denial – but he wanted science to be secular. God was not to enter its discussions. Evolution was adopted not in its own right, but as a banner for the nonreligious, naturalistic science that Huxley (and friends) were pushing. Tellingly, in his 150-hour-long lecture-course on biology, evolution would get about thirty minutes. (Natural selection, ten minutes!) On being questioned about this by a curious student, Huxley responded that evolution had no place in regular science. It should rather be restricted to public lectures, pushing his overall vision of science (Ruse 1996, 43).

What about causes? What about natural selection? With an interesting exception to be mentioned later, Huxley never accepted it as a proven explanation for evolution: "there is no positive evidence, at present, that any group of animals has, by variation and selective breeding, given rise to another group which was, even in the least degree, infertile with the first" (Huxley 1860, 74–75). With bulldogs like this, Darwin might as well have had a Yorkie, friendly without pretentions of being ferocious. Although, to be frank, when it came to causes, Huxley was not much of a dog of any breed. Sometimes, thoughts of jumps (saltations) are more or less endorsed. Generally, however, Huxley's thoughts were elsewhere. Given that Huxley's initiation into the world of biology was through embryology, one suspects that his overall position was close to that of the Romantics. The main point was first that the science Huxley did – and particularly in the early years he did a great deal – was more descriptive than causal.

Causal Problems

What about someone who, like Darwin, wanted to be causal? No one was going to deny that there is a struggle for existence and, more importantly, a struggle for reproduction. Malthus had left his mark. Few were going to deny that some will get through and some will not, that there will be differences between the successful and the unsuccessful. Differences between the fit or fitter and the unfit or less fit. The question was whether, even if natural selection were working flat out, it could have the effects that Darwin claimed it would, namely ongoing systematic change in the direction of ever-improved adaptations, in groups of organisms reproductively distinct. For someone such as Huxley, who was uninterested in causes, it was a question that could be brushed aside. For a Newtonian such as Darwin, who was obsessed with causes, it was a question not to be ignored. Or, to be a little more precise, it was a question that should not be ignored. Darwin may have used only pretend examples. A Darwinian should go beyond this. What about the variations on which natural selection supposedly works? Even after that massive study of barnacles, Darwin admitted bluntly: "Our ignorance of the laws of variation is profound" (Darwin 1859, 167). We don't know what causes variations, how frequent they are, what different kinds they are, and crucially what happens to variations during reproduction. Is the pattern of human sexuality the norm? Males and females give birth to males and females. Or is the pattern of human skin color the norm? One white and one black spouse produce intermediate kids. If the latter, and it really does seem that this is the common pattern, then however effective natural selection is in one generation, its results will be blended out in two or three generations.

The Scottish engineer (and former classmate of Darwin) Fleeming Jenkin made the point with a paradigmatic Victorian example. He supposed that a white man was shipwrecked on an island whose inhabitants were black people. The white man (naturally) becomes king and has lots of offspring: "In the first generation there will be some dozens of intelligent young mulattoes, much superior in average intelligence to the negroes. We might expect the throne for some generations to be occupied by a more or less yellow king; but can any one believe that the whole island will gradually acquire a white, or even a yellow population…?" (Jenkin 1867, 289–90). Selection is "powerless to perpetuate the new variety."

In the fifth edition of the *Origin*, Darwin responded to this "able and valuable" article by agreeing that one or two variations would not do the job, but that, if there were many variations, selection could have effects. "If, for instance, a bird of some kind could procure its food more easily by having its beak curved, and if one were born with its beak strongly curved, and which consequently flourished, nevertheless there would be a very poor chance of this one individual perpetuating its kind to the exclusion of the common form." However, "there can hardly be a doubt, judging by what we see taking place under domestication, that this result [of the individual perpetuating its kind] would follow from the preservation during many generations of a large number of individuals with more or less curved beaks, and from the destruction of a still larger number with the straightest beaks" (Darwin 1869, 5th edition, 104–5).

This was all a little bit ad hoc, even though Darwin tried to give some theoretical backing to his assumption by introducing his "provisional hypothesis" of "pangenesis." Supposedly there are little gemmules all over the body, which get altered by external circumstances, then circulate through the body being collected in the sex cells, thus causing abundant variation in the next generation. No one was much impressed by this. Darwin himself, although he introduced it in his two-volume *Variation of Animals and Plants under Domestication* (1868), kept it out of the *Origin* as well as the later *Descent of Man* (1871). In any case, it seems more a mechanism for supporting Lamarckism – the gemmules get altered as the blacksmith's arms get bigger and more muscular – than something speaking directly to the swamping problem of Fleeming Jenkin. There was an unsolved difficulty at the heart of Darwin's theory, and scientists knew this.

The swamping problem came from within. There was another problem, of a somewhat different ilk, that came from without (Burchfield 1975). "I am greatly troubled at the short duration of the world according to Sir W. Thomson, for I require for my theoretical views a very long period before the Cambrian formation" (Letter to James Croll, January 31, 1869). Darwin is referring to the fact that the physicist William Thomson, later Lord Kelvin, one of the leading scientists of the day, had used physical data – essentially treating the Earth as a cooling body that started as a mass of molten rock, estimating temperature gradients and so forth – to calculate the age of the Earth at a maximum of 20 to 400 million years with 98 million as the best estimate (Thomson 1869). A long time, but hardly long enough for the slow process of Darwinian evolution through natural selection. In the first edition of the *Origin*, Darwin calculated that eroding the area between the North and South Downs, chalky areas below London – "denudation of the Weald" – would have taken 300 million years. Later, under criticism, he withdrew that estimate, but it gives an idea of the difference between the kind of time span that the evolutionist Darwin needed – the denudation would be just one small period in a total age of literally billions of years – and the kind of time span that the physicists were prepared to allow him.

Over the years, Darwin and his fellow biologists and geologists wriggled under the limits supposedly set by physics. Huxley, perhaps expectedly, brushed off the problem: "Biology takes her time from geology" (Huxley 1900, 331). Whatever the causes, they can speed things up or down as is needed. Others were more concerned. For instance, at the end of the 1870s, we find Charles Darwin rejoicing at the suggestion by his physicist son George that the tides caused heating through friction and so slowed the cooling of the Earth, thus throwing Kelvin's calculations off balance. Truly, though, matters rested until the beginning of the twentieth century, when the discovery of radioactive decay and its warming effects showed that Kelvin's calculations were totally off the mark. Today, we think the universe is 13.8 billion years old, the Earth 4.5 billion years old, and the start of life relatively soon after that (3.8 billion years ago). There is quite enough time for natural selection to do its job. One should add that Kelvin lived to see his calculations overthrown, a result he accepted with more or less good grace, although he never publicly announced his retraction.

The age of the Earth question did get solved. The variation problem really did not get solved. We shall pick up on this in the next chapter.

America

Asa Gray, a botanist at Harvard and a close friend of Darwin – he had been let in on the natural selection secret a year or two before the *Origin* (part of a letter from Darwin to Gray was read at the July 1858 Linnean Society meeting) – was empathetic to the Darwinian position, but he was such a committed Christian he had difficulty accepting the nontheistic-mechanist picture he thought was implied by the *Origin*. For the general community, one might have predicted that it was in America that Spencer had his real influence, an influence we shall learn has lasted right down to the present. From the first, without exception, Americans thought of evolution in a Romantic mode. The reason is simple. Hugely influential was Louis Agassiz, he who blew up Darwin's thinking about the parallel roads of Glen Roy. He emigrated to the New World, where he became a professor at Harvard and founded the Museum of Comparative Zoology. Agassiz was an enthusiastic student of Schelling: "a man can hardly hear twice in his life a course of lectures so powerful as those Schelling is now giving on the philosophy of revelation" (a schoolmate's words, quoted in Agassiz 1885, 116). Like Haeckel, we find him preaching a kind of biogenetic law. He affirmed "that the successive creations have undergone phases of development analogous to those of the embryo in its growth and similar to the gradations shown by the present creation in the ascending series, which it presents as a whole" (Agassiz 1833, 24).

Agassiz had also been a student of Georges Cuvier, and that perhaps accounts for the fact that he himself could never accept evolution, even though all his students – including his son Alexander – became enthusiastic evolutionists. Much of their emphasis was on embryology and morphology, supplemented by paleontology – this was the time when those fabulous fossil deposits of the American West were being uncovered – and so, as one could anticipate, they became evolutionists of a Romantic flavor: lots of progress from the past to the present and little, if any, interest in natural selection. A good example is the already-encountered Alphaeus Hyatt, Harvard graduate and student of Agassiz, whose teaching "produced a profound effect upon young Hyatt's mind" (Ruse 1996, 251). As a good Romantic, Hyatt believed strongly in progress, a belief that sparked the already-quoted response by Darwin that he did not believe in any such necessary tendency to progress. (Hyatt was a man much loved. Although they differed, the correspondence between the men is remarkably warm and respectful.) Hyatt did, however, add his own twist about old age and degeneration. As individuals show signs of aging, so also do evolving lines show such signs of aging: "terminal forms are at the same time the highest of their series in their organization and development and yet like the most immature in many characteristics. Again these terminal forms have not only these resemblances but they also resemble the old age of earlier species of their own series" (Darwin 1985: Letter from Hyatt, November 1872).

As the nineteenth century moved towards its end, increasingly it became apparent – with slums and unemployment and arms races between nations (especially with Germany) – that optimistic claims about progress are unwarranted. Visiting Italy as a teenager, Hyatt write: "The lazzarone live, beg, starve, make love and shit upon the church steps and along the quay, which last being the most public is the place generally preferred for the last picturesque action" (Ruse 1996, 255, quoting from Hyatt's unpublished *Travel Book*, 6). Beliefs in degeneration were reflected in the science. E. Ray Lankester, a younger biologist in Britain, was likewise much given to such gloomy thoughts, although what spurred him were less the defecating habits of foreigners and more a sense of personal sexual inadequacy, brought on by all-male education, which led him unable to engage with women of his own social class and hence the need of frequent visits to the brothels of Paris (Ruse 1996, 227). In these directions, Hyatt had suspicions of his own. Coeducation has predictable consequences since "women would be tending to become virilified and men to become effeminized, and both would have, therefore, entered upon the retrogressive period of their evolution" (Hyatt 1897, 91). As always, natural selection generated little enthusiasm: "For years past I have been of the opinion that the regularity with which these series followed out a given line of characteristics in their progress was irreconcilable with your law" (Darwin 1985: Letter from Hyatt to Darwin, 1875).

The pattern was set, and it continued strongly into the twentieth century. Evolution was all "saltations" (jumps from one form to another) or some kind of internally directed evolution (often known as "orthogenesis").

Typical was the (already-mentioned) Stanford entomologist Vernon Lyman Kellogg. In Kellogg's *Darwinism Today* (published in 1905), one is flagged at once that things do not bode well for the old naturalist from Down House. The first chapter – "Introductory: The 'Death-Bed of Darwinism'" – sets the tone. Natural selection is chiefly a kind of clean-up process after the important work has been done by new variation: "The living stream of descent finds its never-failing primal source in ever-appearing variations; the eternal flux of Nature, coupled with this inevitable primal variation, compels the stream to keep always in motion, and selection guides it along the ways of least resistance" (Kellogg 1905, 374). Hardly A+ quality work. "Darwinism, then, as the natural selection of the fit, the final arbiter in descent control, stands unscathed, clear and high above the obscuring cloud of battle. At least, so it seems to me. But Darwinism, as the all-sufficient or even most important causo-mechanical factor in species-forming and hence as the sufficient explanation of descent, is discredited and cast down."

Anticipating later discussion, apparently, worries that there was no genuine *Darwinian* revolution are well taken. Evolution is accepted. Natural selection might even have a clean-up role. But it simply isn't a major cause – let alone *the* major cause – of organic evolution.

Natural Selection at Work

Before we close up shop and go out for a beer, let us pause for a moment and ask whether this is – whether this can be – all that is said on the matter. Thus far, we have been looking at life scientists whose work is not immediately dependent on finding causes – at least, not evolutionary causes. This is not to deny that someone studying the fossil record is going to wonder how it all came about. Why, for instance, to use a classic example, did the horse evolve from a five-toed animal – *Eohippus* – to an animal with a single hoof? The horse paleontologist is going to have suspicions about how some possible processes might have worked – or not. If, for instance, one sees regression in the record – five toes, three toes, four toes, one toe, back to five toes, one is going to wonder about claims for ongoing progress-causing processes (Figure 4.4). Likewise, if there seems no change in the environment, then claims about natural selection seem less plausible. And indeed, as generations of anti-evolutionists – such as Louis Agassiz – show, evolution is not really that important for the science. People such as T. H. Huxley confirm this. In 1890, Hyatt wrote a textbook in which he warned: "We strongly advise teachers not to use this or any theory in teaching minds" (Ruse 1996, 276). Outside the classroom, the *Origin* had a major role. Inside the classroom, it did not.

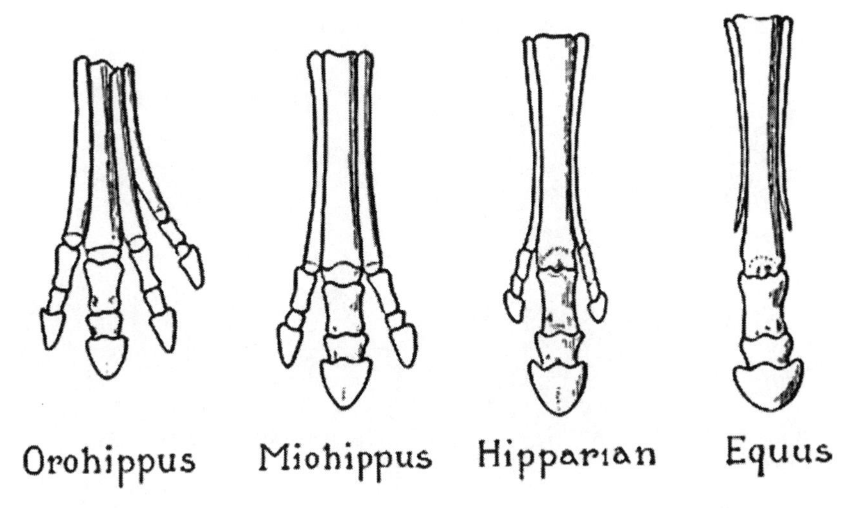

Figure 4.4 The evolution of the toes of horses (T. H. Huxley, *American Addresses*, 1877).

Were there no life scientists for whom natural selection would be an indispensable tool? There were: Alfred Russel Wallace for a start. At the level of science – more on this qualification later – he was always a natural selection booster. Some may doubt this enthusiasm:

> But we claim for Darwin that he is the Newton of natural history, and that, just so surely as that the discovery and demonstration by Newton of the law of gravitation established order in place of chaos and laid a sure foundation for all future study of the starry heavens, so surely has Darwin, by his discovery of the law of natural selection and his demonstration of the great principle of the preservation of useful variations in the struggle for life, not only thrown a flood of light on the process of development of the whole organic world, but also established a firm foundation for all future study of nature. (Wallace 1889, 9)

Wallace was not much of an experimentalist, but he worked and theorized all over the place, especially in the field of biogeography. The dividing line points to different organisms of different origins (Figure 4.5).

If you want to study natural selection in action, you turn to fruit flies, not elephants. What, then, of those working on fast-breeding organisms? Henry Walter Bates, a naturalist/collector, who went off to the Amazon with Wallace, set the pace early, being deservedly famous for his studies of mimicry in butterflies (Figure 4.6). Just after the *Origin* was published, Bates came up with a Darwinian explanation, giving early evidence of the

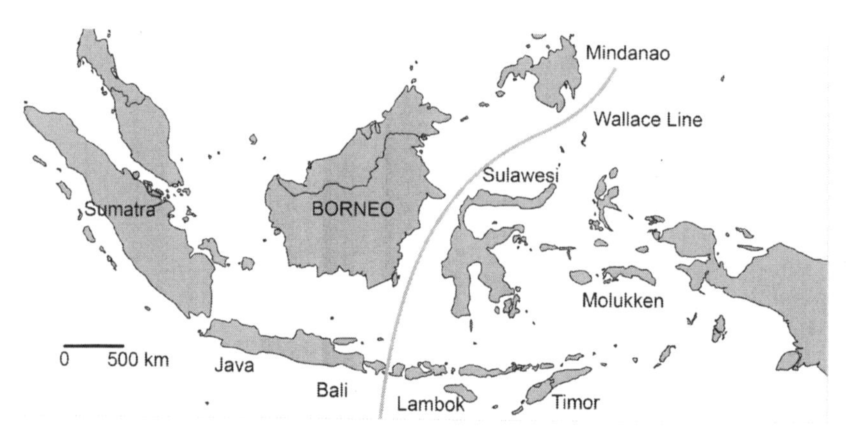

Figure 4.5 "Wallace's Line," showing the abrupt break between the kinds of organisms to the left (west) and the kinds to the right (east).

Figure 4.6 Batesian mimicry. The butterfly to the top is mimicking the butterfly to the bottom. Bates, *Transactions of the Linnean Society* (1862).

power of natural selection. Mimicry, he showed, must have a reason. One starts with the fact that the usual form of such mimicry is where nonpoisonous butterflies are mimicking poisonous forms, even though there is no close relationship. Bates showed that this is a function of natural selection, with the nonpoisonous forms piggybacking, as it were, on the poisonous butterflies (Bates 1862). He told of a kind of butterfly, Leptalis, such that "when a variety arose which happened to resemble any common butterfly inhabiting the same district (whether or not that butterfly be a variety or a so-called distinct species) then that this one variety of the Lactalis had from its resemblance to a flourishing and little persecuted kind a better chance of escaping destruction from predacious birds and insects, and was consequently oftener preserved..." The consequence? "[T]he less perfect degrees of resemblance being generation after generation eliminated, and only the others left to propagate their kind" (Ruse 1996, 223). In other words, natural selection!

Others did similar work. The German-born naturalist Fritz Müller, also working in Brazil, came up with another selection-based explanation of mimicry: Different species of poisonous butterflies grow to look alike, so that predators more quickly learn that that form is not for good eating. He tells us: "Hyridia and Ituna both belong to the class of cases in which the two species which resemble one another appear to be equally well protected by distastefulness." What is the message one extracts? If the predators know by instinct about the distastefulness of the butterflies, then nothing. "But if each single bird has to learn this distinction by experience, a certain number of distasteful butterflies must also fall victims to the inexperience of the young enemies." Suppose now that two species, both distasteful, "are sufficiently alike to be mistaken for one another, the experience acquired at the expense of one of them will likewise benefit the other; both species together will only have to contribute the same number of victims which each of them would have to furnish if they were different." In other words, each species, if equally common, "will derive the same benefit from their resemblance – each will save half the number of victims which it has to furnish to the inexperience of its foes" (Müller 1879, xxvii).

Evolutionists refer to these two (noncompeting) processes as Batesian and Müllerian mimicry. These are the best-known early cases. As the years went by, there was a steady stream of people using selection as the important cause of change. The German biologist August Weismann, today celebrated for his ferocious attacks in the 1880s on Lamarckian inheritance, made his philosophy crystal clear: "I deny the possibility or conceivability of the contemporaneous co-operation of teleological and of causal forces in producing any effect," rather "I maintain that a purely mechanical conception of the processes of nature is alone justifiable" (Weismann 1882, 688). Working within the mechanical perspective, he reported on a major study on the markings of caterpillars:

> It has been possible to show that each of the three chief elements of the markings of the Sphingidæ [sphinx moths] have a biological significance, and their origin by means of natural selection has thus been made to appear probable. It has further been possible to show that the first rudiments of these markings must also have been of use; and it thus appears to me that their origin by means of natural selection has been proved to demonstration. (Weismann 1882, 380)

There was no ambiguity about the cause or what deserved credit: "[I]t has been established that each of the elements of marking occurring in the larvae of the Sphingidae originally possessed a decided biological significance, which was produced by natural selection" (Weismann 1882, 388).

In the next generation, Edward B. Poulton (1890, 1908), professor of zoology at Oxford, made significant selection-based contributions to problems of animal coloration. Expectedly, from one interested in birds, sexual selection is strongly endorsed. "When we look at the marvellous eyes upon the train of a Peacock, or the more beautiful markings on the feathers of the male Argus Pheasant, it seems impossible that so wonderful and complete a result can have been produced by the aesthetic preferences of female birds." But Darwin shows that it can be and is so produced: "the relation between these characters and much simpler markings on other parts of the surface. He proves that the one has been derived from the other by gradual modification … Such facts, while eminently suggestive of … some selective agency, seem to be unexplained by any other theory" (Poulton 1890, 334).

At the end of the century, turning to marine invertebrates, W. F. R. Weldon (1898) began doing groundbreaking selection-based studies on crab adaptations. In this he was supported by the mathematical genius of Karl Pearson (1900) – who made his name as a statistician devising the chi-squared test, among other innovations. Thanks to long-term study, Weldon found that the frontal breadth of shore crabs around the marine station in Plymouth (on the English south coast) was getting ever smaller. He linked this to increased silt in the water, a result of a new breakwater built across Plymouth Sound. Weldon hypothesized that the change of breadth was an adaptation to deal with the increased silt, and in a dazzling series of experiments he showed that crabs in water with extra silt evolved towards smaller frontal breadths and crabs in clean water went the other way, towards greater frontal breadths. His triumphant conclusion: "I hope I have convinced you that the law of chance enables one to express easily and simply the frequency of variations among animals, and I hope I have convinced you that the action of natural selection upon such fortuitous variations can be experimentally measured, at least in the only case in which any one has attempted to measure it." He added: "I hope I have convinced you that the process of evolution is sometimes so rapid that it can be observed in the space of a very few years" (Weldon 1899, 141).

Darwin was not unappreciative of this line of work. He managed to get the lower-middle-class Bates a good job as secretary of the Royal Geographical Society and he wrote a fulsome foreword to the translation of Weismann's work into English. Notwithstanding, these selection-based studies did not make an overwhelming impression. In major part, Darwin himself was to blame for this. He was never overly convinced that selection could be seen in action today. With his money, he could easily have set up a selection institute in his back garden or at Kew or somewhere similar. All too predictable is that, when a Scotsman wrote to Darwin and suggested that he come down to England and set up just such an institute, Darwin put him off by getting him a job in India! "If my health had been better I would have proposed to you to have come here and have worked for a couple of years on scientific subjects, but at present, and probably for ever, this is impossible" (to John Scott, April 9, 1864). This attitude clearly influenced Darwin's attitude to the selection-work of others. The job for Bates was a little bit of a poisoned chalice. Instead of being a full-time Darwinian biologist, he became the servant of a society of upper-middle-class toffs, catering to their needs. Although Darwin mentioned Bates's work in later editions of the *Origin*, these mentions came towards the end of the book – in the grab-bag penultimate chapter along with systematics and so forth – rather than (in Chapter IV) where selection is introduced and discussed.

One significant fact is that much of the selection work was being done by amateurs and enthusiasts rather than professional scientists. Properly inspired, directed, and appreciated, these amateurs could have been the fount of huge amounts of pertinent information. For instance, there was considerable interest in what is known as "industrial melanism," where moths and butterflies evolve in ways better to camouflage themselves against the barks of trees getting ever darker from the soot produced by industry (see Figure 5.1). "I believe … that Lancashire and Yorkshire melanism is the result of the combined action of the 'smoke,' etc., plus humidity [thus making bark darker], and that the intensity of Yorkshire and Lancashire melanism produced by humidity and smoke, is intensified by 'natural selection' and 'hereditary tendency'" (Tutt 1890, 56). There are times when the indifference of professional scientists towards this work by amateurs is nigh-on heartbreaking. To illustrate. In 1878, Albert Brydges Farn, a civil servant and keen sportsman (that is, killer of birds), wrote the most remarkable letter to Darwin:

My dear Sir,

The belief that I am about to relate something which may be of interest to you, must be my excuse for troubling you with a letter.

Perhaps among the whole of the British Lepidoptera, no species varies more, according to the locality in which it is found, than does that Geometer, Gnophos obscurata. They are almost black on the New Forest peat; grey on limestone; almost white on the chalk near Lewes; and brown on clay, and on the red soil of Herefordshire.

Do these variations point to the "survival of the fittest"? I think so. It was, therefore, with some surprise that I took specimens as dark as any of those in the New Forest on a chalk slope; and I have pondered for a solution. Can this be it?

It is a curious fact, in connexion with these dark specimens, that for the last quarter of a century the chalk slope, on which they occur, has been swept by volumes of black smoke from some lime-kilns situated at the bottom: the herbage, although growing luxuriantly, is blackened by it.

I am told, too, that the very light specimens are now much less common at Lewes than formerly, and that, for some few years, lime-kilns have been in use there.

These are the facts I desire to bring to your notice.

I am, Dear Sir, Yours very faithfully,

A. B. Farn

(Letter from Albert Brydges Farn on November 18, 1878 [Darwin 1985, 26, 440])

Darwin should have brought out a new edition of the *Origin* with this letter printed opposite the title page. Far from it. Darwin seems not to have replied to the letter! Is one amazed that this selection-based strain of evolutionary thinking has escaped public notice?

And into the Twentieth Century

There were issues, important issues, that were raised by the *Origin*, and not all of them were tackled successfully in the years following its publication. Nevertheless, as we turn now to the twentieth century and the scientific fate of the theory of evolution by natural selection, one final point should be made, or at least restressed. We have seen how Huxley and company were essentially doing science that was not causal, or at least science where finding and using causes was not top priority. More positively, what we have seen is how Huxley and company used the prestige of Darwin and his work to support their own drive for professionalism – secular professionalism. A man who had written one of the most popular travel books of the mid nineteenth century, a dedicated researcher who spent eight years working on barnacles, and now who had written the best book by far on the origin of organisms, whether or not one accepted this in full – this was no *Vestiges* – this was a man of genuine scientific research, a family man of moral worth who battled on despite horrendous ongoing sickness. Darwin gave these strivers for secular professional science an aura of quality – an aura that they got from no other. Moreover, never forget that it was Darwin who, through his consilience, gathering evidence from the whole of biology – behavior, fossils, geographical distributions, classification, morphology, embryology, rudimentary organs – made evolution a fact and not merely some flaky speculation. Evolution was here to stay. So here, in a very real sense, it was Charles Darwin's revolution, not that of Herbert Spencer or Richard Owen or Louis Agassiz. Charles Darwin's revolution with respect to *the fact of evolution* in itself, as well as his revolution with respect to *the mechanism of natural selection*. Expectedly, others did talk about different areas of biology being subsumed under the evolutionary hypothesis. Looking at two as examples, Chambers talked somewhat unsystematically about fossils and not a lot about anything else (and when he does it can be difficult to see his point), and Spencer's *Principles of Biology*, published five years after the *Origin*, has over 140 references to the *Origin* (more in later editions). At times one wonders who is writing the book. Darwin wins the prize! Indicative is that, when he died in 1882, there was to be no nonsense about his being buried in the local church cemetery. They wanted him buried in England's Valhalla, Westminster Abbey, and so he was, along with the only other English scientist of his status, Isaac Newton.

5

Evolution in the Twentieth Century

In chapter 1, much was made of the differences between and implications of the two root metaphors: the world as organism and the world as machine. As we continue our story, digging into the history of Darwinian thinking in the twentieth century, these two metaphors continue to be very important. We shall see them structuring the discussions, especially those discussions where there are points of violent disagreement. This is not to deny that, as we move forward, the presumption is going to be that the operative metaphor is, as for the physical sciences, mechanism. This certainly holds true for the first part of our ongoing history, so without constantly bringing it to the fore, let it simply be noted that for now it is the operative root metaphor.

Gregor Mendel

The elephant in the room: Darwin's theory was incomplete. If and when the theory was completed, could natural selection be that effective? After his eight-year study of barnacles, Darwin knew a lot about variation. He knew that it was always there in natural populations. He knew that it rarely contributed directly to the adaptive efficiency of the organisms in which it occurred. In this sense, it was "random," meaning not "without cause" but "without direct utility." It was on this variation that selection worked. Selection was the cause of change – adaptive change. This said, Darwin knew little or nothing about the systematic nature of variation or how it was caused. We have seen that he proposed the not very successful hypothesis of pangenesis, but obviously he himself thought little of it. We saw it did not find its way into the *Descent of Man*. And with its central commitment to Lamarckian change, for all that Herbert Spencer remained an unflagging enthusiast for such change – the inheritance of acquired characteristics – before the century was out, more and more doubt was being cast on the supposed process. Indeed, back when he floated the idea of pangenesis, Darwin himself had doubts. "With respect to Jews, I have been assured by three medical men of the Jewish faith that circumcision, which has been practised for so many ages, has produced no inherited effect" (Darwin 1868, 2, 23).

Before the century was out, the stake seems to have been driven through the heart of Lamarckism. German biologist August Weismann, mentioned in Chapter 4, proposed his "germ plasm" theory, arguing that heritable information occurs only in the reproductive units and not in the physical (somatic) body. Against Lamarckism, he showed that repeated amputation of the tails of mice made no difference to the appearance and growth of tails in subsequent generations: "901 young were produced by five generations of artificially mutilated parents, and yet there was not a single example of a rudimentary tail or of any other abnormality in this organ" (Weismann 1889, 432). Critics have argued that this was not a fair test of Lamarckism, which claims that changes come from organisms themselves acquiring new features – the longer neck of the giraffe – rather than externally imposed changes, with no obvious adaptive value – amputating the tails of mice. However, strictly relevant or not, thinking such as Weismann's spelt the decline of faith in Lamarckism if not its absolute refutation.

In any case, by century's end, the whole Lamarckism debate was pushed to one side by the arrival of a new line of thought: "Mendelism." We know already that a major criticism of Darwin's theory (as made by people such as Fleeming Jenkin) was the general belief that characteristics would, on balance, be blended from generation to generation. As we all now know, the solution to Darwin's problem was being discovered across Europe, in what then was the Austro-Hungarian Empire, by the friar Gregor Mendel (Bowler 1989). Discovered, but one of those not-uncommon cases where the discoverer does not see the incredible implications of what he was finding. Mendel had a copy of the *Origin* (in German), but although he admired Darwin's work, Mendel was always a little cagy about total commitment. A collaborator and friend gave what seems to be a balanced view, that Mendel "who was greatly interested in the idea of evolution, was far from being an adversary of the Darwinian theory, but always when Darwin's name came up, he said that the theory was inadequate, that something was lacking" (Fairbanks 2020, quoting Gustav Niessl von Mayendorf).

"Something was lacking"! One wonders what would be the reaction of Mendel were he put into a time machine and carried into the twentieth century, and found that this "something" was being provided by the work of an unknown friar in the Austro-Hungarian Empire of around 1860? Would he be proud and pleased or would he feel like the biggest fool of all time? There are few of us who have not had the second emotion at some point in their lives, but to be fair there are even fewer who have brought it on in quite such spectacular fashion. It was at the beginning of the twentieth century – by which time there was much greater knowledge of the physical nature of the cell – that three people, separately, realized the significance of Mendel's work, and then things started to move forward rapidly. The work of Mendel on pea plants provided the empirical basis for the claim – hypothesis – that physical characteristics such as flower color are controlled by "factors" – units of inheritance, what we now call "genes." In sexual organisms, these factors come in pairs. Factors that are or can be paired are called "alleles" (or, more cumbersomely, "allelomorphs"). Central to the hypothesis is the claim that which factor (allele) is transmitted to an offspring comes entirely by chance and is quite independent of the other member of the pair. (Added is the claim that what happens with one set of alleles is independent of what happens with other sets.) Moreover, part of the hypothesis is the additional claim that the effects of one factor generally block out the effects of the other factor – "dominant" and "recessive." All important, as can be seen from following box, is the fact that there is no blending of features out of existence. The factors, the genes, go on unchanged, and this means that the physical characteristics can keep reappearing unchanged. These physical characteristics are known today as the "phenotype"; as opposed to the genetic characteristics, the "genotype."

MENDEL'S FIRST LAW

One parent has two grey-causing alleles and the other two white-causing alleles. If grey is dominant over white, the offspring are grey. But then, among their offspring, white organisms will reappear. (The ratio in F2, generation two, will be 3:1 grey over white, because, on average, you will get one plant with two grey-causing alleles, hence physically grey; two plants with one grey and one white, hence physically grey; and one with two white, hence physically white. Complexifying things a little, sometimes neither factor is dominant and the phenotype is intermediate. The causal process is unchanged, but the F1 ratio will be all pale grey, and the F2 ratio will be one grey, two pale grey, and one white.)

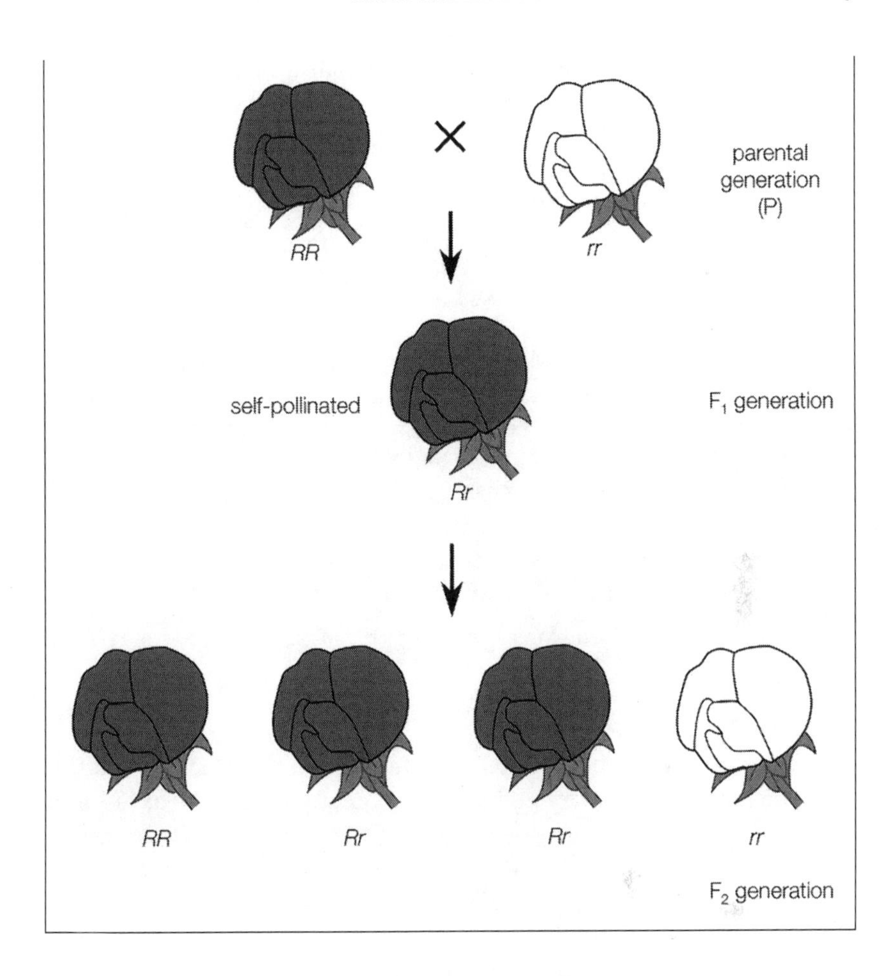

In 1894, the English biologist William Bateson published *Materials for the Study of Variation, Treated with Especial Regard to Discontinuity in the Origin of Species* (Bateson 1894). The title announced his interest in the nature of variations, and his subtitle hinted at his unease with the supposed continuous change in Darwinian selection theory. Mendelism was the answer to the dream of a man who increasingly felt out of step with Darwinian enthusiasts, not the least being the selection-proselytizing Weldon – who, incidentally, had been Bateson's teacher. Oedipal issues? Whatever the underlying psychology, Bateson at once picked up on the new Mendelian approach to heredity and became its champion. Against the professional supporters of natural selection, Bateson presented Mendelism as a contradicting alternative, or, more accurately, a cutting-down-to-size alternative. "The reader who has the patience to examine Professor Weldon's array of objections will find that almost all are dispelled by no more elaborate process than a reference to the original records" (Bateson 1902, xi). This was one of the nicer things said as, quick off the mark, Bateson pushed his vision in *Mendel's Principles of Heredity: A Defence.* Mendel's achievement should be appreciated for its worth, especially when compared with other factors in evolutionary change. Spelling things out, coming in apparently as a friend, and then the knife in the gut:

> There is also nothing in Mendelian discovery which runs counter to the cardinal doctrine that species have arisen "by means of Natural Selection, or the preservation of favoured races in the struggle for life," to use the definition of that doctrine inscribed on the title of the Origin. By the arbitrament of Natural Selection all must succeed or fail. Nevertheless the result of modern inquiry has unquestionably been to deprive that principle of those supernatural attributes with which it has sometimes been invested. The scope of Natural Selection is closely limited by the laws of variation. (289)

"Clean-up operation" is the most-generous phrase that comes to mind.

And So to Population Genetics

Almost necessarily the history of science tends to be sanitized. Every interesting claim seems to have multiple exceptions, and Mendel – as given in the basic, stripped-down version of the previous section – fits this pattern. Weldon did not so much reject Mendel as show, starting with peas, that there are many cases where Mendel seems inapplicable. The initial rancor faded somewhat, more because of Weldon's sudden death in 1906 than any meeting of minds. Mendel became the touchstone against which problems and exceptions could be measured and the reasons for them could be explored and explained. "For the Mendelian, the interesting challenge now was not to show that there were departures from simplicity – that was admitted – but to figure out how to modify the principles so as to explain those departures" (Radick 2023, 189). So great were some of the modifications that sometimes one feels that retention of the name of "Mendel" was more a function of the historical authority conveyed than any genuine continuity – as Weldon would have hurried to highlight. But, in our post-Kuhnian world, we realize that scientists are not always entirely rational in promoting their positions. Winning over the reader is the true aim. Deservedly or not, the name of Mendel has stuck.

Important for our story is that things did not stand still (Provine 1971). Though this was not work done directly to cement the place of natural selection, crucially significant was the unraveling of the nature of the units of heredity, the "genes." This was the work, at Columbia University in the second decade of the century, by Thomas Hunt Morgan and his associates (Allen 1978). They showed that genes are physical things, to be found along the chromosomes, string-like entities in the centers, nuclei, of complex cells. Underpinning Mendel's hypothesis, they showed also that the chromosomes are paired, so genes are twinned, alleles, with the possibility of identical pairs – "homozygotes" – or different pairs – "heterozygotes." Genes can be altered, spontaneously as it were – "mutations" – and there is no reason to believe that their new nature is particularly adaptively helpful to their possessors. What is crucially important is that mutations can have very small effects which can be combined in an additive manner. Mendelian genetics is far from incompatible with gradual, physical – phenotypic – variation.

In other words, the position now seems equally supportive of a gradualist such as Weldon. He was just not around to promote his perspective (Radick 2023, 362–63). Either way, whatever you call it, this did not mean that people now stopped thinking in a Mendelian fashion. One could recognize the importance of Morgan's work, without feeling it necessary to translate everything into his physical terms and dropping the more theoretical "Mendelian" gene or factor. Morgan was more a support than a rival replacement. As a man about to play a leading role in our story, the eminent statistician Ronald Fisher, put things in 1930: "The apparent blending in colour in crosses between white races of man and negroes is compatible with the view that these races differ in several Mendelian factors, affecting the pigmentation" (Fisher 1930, 17).

Along with the physical work of Morgan and associates, theoretical mathematics played a significant role. Natural selection is a population-based causal process, as opposed to Lamarckism (now less and less plausible), which is an individual-based causal process. Crucial therefore was the extension of genetics to populations. The all-important breakthrough came in 1908, thanks to the independent work of two thinkers, in England the mathematician G. H. Hardy and in Germany the physician Wilhelm Weinberg. What came to be known as the Hardy–Weinberg law showed that, without external disturbing factors, the genes (alleles) within a population will immediately move to a state of equilibrium. Against the assumption – one clearly behind the critique of Fleeming Jenkin – that minority forms will move to elimination and nonexistence in a population, in large populations such forms will continue to exist, perpetually. An equilibrium law against which – just like Newton's first law of motion – disturbing factors can now be introduced and their effects measured and studied.

HARDY–WEINBERG EQUILIBRIUM

Assume that we have a (large) population, with just two alleles, A and a, in ratio p:q. Hence, $p + q = 1$.

Given random mating, the Hardy–Weinberg law states that, in subsequent generations, the ratio of alleles will remain constant, and the ratio of genotypes will be:

$$p^2AA + 2pqAa + q^2aa$$

The Equations

$$p + q = 1$$

$$p^2 + 2pq + q^2 = 1$$

- A gene has two alleles, A and a
- The frequency of allele A is represented by p
- The frequency of allele a is represented by q
- The frequency of genotype AA = p^2
- The frequency of genotype aa = q^2
- The frequency of genotype Aa = 2pq

Early Mendelians such as Bateson, eager to promote their new thinking, tended to present the Darwin–Mendel confrontation as a matter of right and wrong. One of the two, Darwinism or Mendelism, was right or the more profitable and the other the loser and less profitable. Weldon versus Bateson. As initial enthusiasms started to decline, a coherent overall "population genetics" picture – in significant respects synthesizing the Darwinian and the Mendelian contributions into one united whole – started to emerge around the end of the third decade of the century, between 1930 and 1932. In the Anglophone world, three men are usually picked out – in Britain Ronald A. Fisher (1930) and J. B. S. Haldane (1932), and in America Sewall Wright (1931, 1932). Here, I will focus on the more influential Fisher and Wright. Building on their work, we have the experimentalists who started to put empirical flesh on the theoretical skeletons, specifically E. B. Ford (1931, 1964) and his school of "ecological genetics" in Britain and the Russian-born Theodosius Dobzhansky (1937) and associates following in the US. The former version of the Darwin–Mendel unification is usually known as "Neo-Darwinism"; the latter as "the synthetic theory" of evolution. We shall take the story as far as

1959, the hundredth anniversary of the *Origin*. This is a somewhat arbitrary marker, but not entirely. It was when evolutionary studies could justifiably be said to have found its "paradigm."

Ronald A. Fisher

The Hardy–Weinberg law points the way to thinking in terms of large populations – large populations of organisms. The law directs us to focus on the genes in those populations ("gene pools"). Without interference by outside factors, in effectively infinite populations, you are going to get equilibrium. "The particulate theory of inheritance resembles the kinetic theory of gases with its perfectly elastic collisions, whereas the blending theory resembles a theory of gases with inelastic collisions, and in which some outside agency is required to be continually at work to keep the particles astir" (Fisher 1930, 11). This means, as Fisher stressed, although mutations are going to be the building blocks of evolution, as it were, this does not imply that it will be mutations that direct the course of evolution. It is natural selection or nothing. As Fisher wrote in 1930:

> The whole group of theories which ascribe to hypothetical physiological mechanisms, controlling the occurrence of mutations, a power of directing the course of evolution, must be set aside, once the blending theory of inheritance is abandoned. The sole surviving theory is that of Natural Selection, and it would appear impossible to avoid the conclusion that if any evolutionary phenomenon appears to be inexplicable on this theory, it must be accepted at present merely as one of the facts which in the present state of knowledge seems inexplicable. (Fisher 1930, 20–21)

One thing stressed by Fisher is that mutations, with but small effects, could in the long run be as effective as mutations with larger effects. "If a change of 1 mm. has selection value, a change of 0.1 mm. will usually have a selection value approximately one-tenth as great, and the change cannot be ignored because we deem it inappreciable." Continuing:

> The rate at which a mutation increases in numbers at the expense of its allelomorph will indeed depend on the selective advantage it confers, but the rate at which a species responds to selection in favour of any increase or decrease of parts depends on the total heritable variance available, and not on whether this is supplied by large or small mutations. (Fisher 1930, 15)

Given Fisher's huge sense of self-worth, it was to be expected that he drew an analogy from physics (kinetic theory of gases), where the (very major) second law of thermodynamics states that entropy always increases (the disorder increases, meaning that the available usable energy is always on the decrease), announcing his equivalent for his version of population genetics. If you have variation in a population, then selection is going to be always at work, pushing the group to the most efficient level of adaptation. Understanding by "fitness" the ability to reproduce compared with that of competitors, and by "variance" the difference from the norm, then Fisher's "Fundamental Theorem of Natural Selection" states: "The rate of increase in fitness of any organism at any time is equal to its genetic variance in fitness at that time." Clearly, it does not follow that a population is always on the up. In a passage redolent of Lyell's uniformitarianism, where there is no absolute progress, Fisher wrote that if "an organism be really in any high degree adapted to the place it fills in its environment, this adaptation will be constantly menaced by any undirected agencies liable to cause changes to either party in the adaptation."

Elaborating, Fisher continued:

> As to the physical environment, geological and climatological changes must always be slowly in progress, and these, though possibly beneficial to some few organisms, must as they continue become harmful to the greater number, for the same reasons as mutations in the organism itself will generally be harmful. For the majority of organisms, therefore, the physical environment may be regarded as constantly deteriorating, whether the climate, for example, is becoming warmer or cooler, moister or drier...

Adding: "Probably more important than the changes in climate will be the evolutionary changes in progress in associated organisms. As each organism increases in fitness, so will its enemies and competitors increase." (41)

Without putting words in people's mouths, one suspects that this would have been a picture with which Charles Darwin would have felt very comfortable. Nothing will happen overnight, and there is a real possibility of reverse – if not to original kinds, then at least to original levels. There is no necessary progress. Cumulatively, it may happen, but no guarantees. Intelligence, for example, might just be too expensive to maintain. (Fisher was an enthusiast for eugenics, so he certainly thought there could be progress. But this would be human-driven, not just the effect of the blind laws of natural selection.)

Ecological Genetics

The evolutionist best known in England was T. H. Huxley's grandson, Julian Huxley (older brother of Aldous). His *Evolution: The Modern Synthesis* (1942) was, as it says, an overview of the field. One doubts that it had major effects on the course of Darwinism. Apart from anything else, rather than straight science, it more often reads as a manifesto to Huxley's conviction that evolutionary thinking's main task is to provide a foundation to build a secular equivalent alternative to Christianity. "One somewhat curious fact emerges from a survey of biological progress as culminating for the evolutionary moment in the dominance of Homo sapiens. It could apparently have pursued no other general course than that which it has historically followed" (Huxley 1942, 559). Indeed!

In the scientific world, complementing Fisher the theoretician, was the Oxford-based naturalist, E. B. "Henry" Ford, who – with his young associates – showed how a Mendel-infused Darwinism could grapple with empirical questions and problems. The problems were often inherited from the nineteenth century. Anyone who lived in Britain in the 1950s will remember the horrendous fogs – "peasoupers" – that would choke cities for days on end. Shades of London in the time of Sherlock Holmes!

> It was a September evening, and not yet seven o'clock, but the day had been a dreary one, and a dense drizzly fog lay low upon the great city. Mud-coloured clouds drooped sadly over the muddy streets. Down the Strand the lamps were but misty splotches of diffused light which threw a feeble circular glimmer upon the slimy pavement. The yellow glare from the shop-windows streamed out into the steamy, vaporous air, and threw a murky, shifting radiance across the crowded thoroughfare. (Conan Doyle 1890, 42)

In the mid twentieth century, people were as enthusiastic as in the nineteenth century about collecting butterflies and moths. Industrial melanism was a fact of life. It was what made the hobby so incredibly interesting – think back to the letter from A. B. Farn to Darwin quoted in Chapter 4 – apart from giving a ready answer to those who disapproved strongly of killing insects in the pursuit of personal pleasure. However, more sophisticated theories and tools could now yield more answers with greater accuracy. Philip M. Sheppard, a student of Ford, discussed melanism carefully in his *Natural Selection and Heredity* published in 1958. He reported on experiments by H. B. D. Kettlewell, who released different colored moths in appropriately different situations. "He found, both by the proportion of the two types recaptured and by direct observation of the birds taking the moths from the trunks, that *carbonaria* was far less heavily predated than the typical form." In one experiment, "of equal numbers of the two forms 43 typical were taken to only 15 *carbonaria*. Consequently, the melanic was at a great advantage, which explains why it has become so common in polluted areas" (72). Conversely, in a nonpolluted area, birds "took 164 *carbonaria* but only 26 typicals when equal numbers of the two varieties were put on tree trunks." In the sixty years and more since

Figure 5.1 Industrial melanism. The tree trunk is black from soot and whereas the dark form of moth is concealed, the light form stands out and attracts predators.

this was written, although almost expectedly there have been doubters and critics, overall the claims have survived the tests of time (Grant 2021). Thanks to clean air laws, the melanic forms have now become less and less common. Trees are far less heavily polluted (Figure 5.1).

New problems were also tackled. Polymorphism, where one has two or more variants in a species, was a rich field for study. In itself, this may seem of no reason for special interest. If, for instance, a new mutant form proves superior to already existing forms, one expects polymorphism as the species changes from one type to the other. More interesting is where selection maintains polymorphism. A much-studied case is "balanced heterozygote superior fitness." Ford (1964) seized on this: "Polymorphism will result if the heterozygote possesses some physiological advantage as opposed to the homozygote; it may for example be more fertile." The most famous recorded instance occurs in humans (Allison 1954a, b). In parts of Africa, and as a consequence among African-Americans today, a proportion of young children are born with a fatal form of anemia. It is now known that there are two alleles involved: the "normal" allele and the "sickle-cell" allele, which in homozygotes causes the red blood cells to collapse into a sickle-type shape, functioning very inefficiently. These latter, up to about 4 percent of the population – Nigeria, for example, has a rate of about 2 percent (WHO 2006) – unless treated by modern medicine die in early childhood. We now know that the reason why the sickle-cell allele persists in populations is that such populations live in areas heavily infested by mosquitoes leading to malaria. Homozygotes for the normal allele are at great risk from malaria. However, heterozygotes for normal and sickle-cell alleles have an immunity to the malaria parasite, to the extent that they are twenty-six times fitter than normal (malaria-unprotected) homozygotes. The sickle-cell anemic children, who will not reproduce, are the cost of keeping the heterozygotes much better protected from the malaria parasite (Gladwin, Kato, and Novelli 2021).

Selection for rareness can also lead to polymorphism. If, for instance, a predator must learn to recognize its prey, then the rare form in a group will be less likely to be predated. But as selection works in its favor, it will increase in number until the predators can recognize it as easily as the other form, and so the two forms will persist in a balance. The snail *Cepaea* has different forms of shell patterns. Thrushes eat these snails, and if one form is at an adaptive advantage – it blends more easily into its background so the thrushes do not see it – this form will increase in number against the other forms. Eventually, it will become so common (and the other forms relatively uncommon) that if the thrushes do not learn to recognize them, they will starve. Hence, there is now going to be selective pressure on the hitherto-advantaged form, and so at some point there will be equilibrium, with none of the forms having an edge over others. (As always, the real world tends to be rather more complex. See Tucker 1994 and Majerus 1998 for overviews.)

Sewall Wright

As we shall see, this work showing how selection can promote variation in populations was to be of major significance. First, we must cross the Atlantic and look at the work and influence of Sewall Wright (Provine 1986). As we do so, we come to one of the most interesting aspects of our whole story, showing that the distinction between the root metaphors of machine and organism is of more than just historical interest (Ruse 1996). As a graduate student Wright went to Harvard, and then was employed by the US Department of Agriculture, where he worked on the breeding and improvement of shorthorn cattle. Always interested in evolution, this picked up when he moved to a lifelong position at the University of Chicago. Fisher and Wright communicated about technical mathematical issues, and they agreed about the mathematics. Indeed, Wright corrected Fisher. However, notwithstanding mathematical agreement, Wright's thinking about evolution led him in a very different direction from that of Fisher – but not at first. He began with the Hardy–Weinberg equilibrium, using this as the background foundation. He wrote in 1931: "The starting point for any discussion of the statistical situation in Mendelian populations is the rather obvious consideration that in an indefinitely large population the relative frequencies of allelomorphic genes remain constant if unaffected by disturbing factors such as mutation, migration, or selection" (Wright 1931, 97).

Wright then introduced the factors that might take a population from equilibrium: mutation, migration, and selection. And it is at this point we start to see his thinking was very different from that of Fisher. Organicism! He stressed the holistic nature of the key Darwinian contribution: "Selection, whether in mortality, mating or fecundity, applies to the organism as a whole and thus to the effects of the entire gene system rather than to single genes. A gene which is more favorable than its allelomorph in one combination may be less favorable in another" (101). Wright agreed with Fisher in that selection can promote, as well as eliminate, diversity. With Fisher and Ford, Wright made mention of heterozygote fitness (more commonly today "heterozygote advantage"), where the heterozygote is fitter than either homozygote, hence keeping different alleles in the same population. "There may be equilibrium between allelomorphs as a result wholly of selection, namely, selection against both homozygotes in favor of the heterozygous type" (102).

Then comes the decisive break, or innovation: "genetic drift" as a function of small populations:

> There remains one factor of the greatest importance in understanding the evolution of a Mendelian system. This is the size of the population. The constancy of gene frequencies in the absence of selection, mutation or migration cannot for example be expected to be absolute in populations of limited size. Merely by chance one or the other of the allelomorphs may be expected to increase its frequency in a given generation and in time the proportions may drift a long way from the original values. (106)

Why is this so very important? The answer comes most clearly in a follow-up paper that Wright wrote for a congress in 1932. First, through what has come to be known as (the metaphor of) an "adaptive landscape," Wright showed how he considered the course of evolution must follow (Figure 5.2). For Fisher, thinking in terms of selection working within large populations, the whole group moves in the direction of what is of the highest adaptive value. As we have seen, Fisher did not think matters were ever that stable – open always to ongoing change, because of environmental changes and so forth.

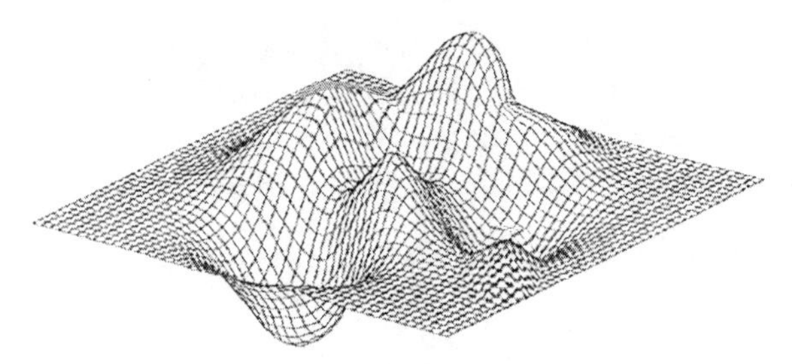

Figure 5.2 Sewall Wright's adaptive landscape. Note that (unlike Fisher's position) the landscape is fixed.

Wright seems to have thought more in terms of stable adaptive "peaks," on which groups find themselves, and the problem now became one of moving to a higher peak. For Fisher, selection does this on its own, for, in a way, peaks – points of highest adaptive value – are being created as the population changes in response to selection. For Wright, between the independently existing peaks are going to be nonadaptive valleys. The problem is how to go down into these valleys so one can then climb up to higher peaks. This is just what selection cannot do.

The peaks of mountains represent areas of high fitness, the valleys areas of low fitness. Suppose a group of organisms were at the top of the lower peak. Say their color is good but not great at camouflage. How could they ever improve their fitness, getting to the top of the higher peak – change their color to something different which is better camouflage – if it means going through changes of color along the way that are not very good at all at camouflage – going down into a valley in order to get to the foothills of the higher peak? Genetic drift can do the job, because it is not constrained by fitness demands. It is not natural selection.

Genetic drift can do the job and *does* do the job! If a large population is divided into many subpopulations, with gene ratios drifting – that is in a non-selection-driven manner – you are going to end up with many subpopulations with different non-adaptive genes fixed in these various subpopulations. They will, as it were, have slipped their populations down from the peak into a valley. Then, rather like individual mutations, one or more will happen to have just what is needed to climb back up to a different higher peak. Either this population spreads its innovative, now-adaptive (for a different peak) genes to other populations, that can then follow up the better peak, or it wipes out the competing now inferior other subpopulations. As Wright wrote in 1932, "The direction is largely random over short periods but adaptive in the long run" (150). Elaborating: "let us consider the case of a large species which is subdivided into many small local races, each breeding largely within itself but occasionally crossbreeding. The field of gene combinations occupied by each of these local races shifts continually in a nonadaptive fashion (except in so far as there are local differences in the conditions of selection)." Continuing:

> With many local races each spreading over a considerable field and moving relatively rapidly in the more general field about the controlling peak, the chances are good that one at least will come under the influence of another peak. If a higher peak, this race will expand in numbers and by crossbreeding with the others will pull the whole species toward the new position. The average adaptiveness of the species thus advances under intergroup selection, an enormously more effective process than intragroup selection. (Wright 1932, 363)

Evolution has occurred! Although, note, hardly Darwinian evolution. The major changes are nonadaptive, coming through drift, and only when the creative work has been done, with one or more populations now ready to climb a new peak, does some form of intergroup selection kick into action. One can start to sense why it was that, while the English called the Darwin–Mendel synthesis by the name "Neo-Darwinism," the Americans did not want it to be thus labeled and the revised theory became known as the "Synthetic Theory." Overall, equilibrium, followed by disruption of equilibrium, followed in turn by achieving a new, higher equilibrium. And if this does not strike a bell, then whatever your age you are having a "senior moment." Pure Herbert Spencer! The crucial creative steps are non-Darwinian and the overall picture is one of dynamic or moving equilibrium.

Can this possibly be so? One of the founders of modern evolutionary thinking was no Darwinian? Someone who discounted the significance of natural selection? An organicist rather than a mechanist? It is true indeed (Ruse 1996). When Sewall Wright was at Harvard, the inspiration of the biology department was Herbert Spencer – hardly meriting comment, given the huge influence Spencer had had on late-nineteenth-century American thinking. Particularly important was the biochemist L. J. Henderson, who in 1917 wrote of Spencer that he was more a visionary than an empirical scientist, but "his generalizations, regarded as provisional and tentative hypotheses, possess genuine importance" (Henderson 1917, 124). Wright was taught by Henderson and fell under his spell. He wrote to his brother Quincy that "I was always very much impressed with Henderson's ideas," and acknowledging explicitly the direct influence back to Spencer: "I found him a very stimulating lecturer and got lots of ideas from him, 'condition of dynamic equilibrium' etc" (Ruse 1996, quoting unpublished letters). Not just this.

As later writings show, the holism also. In Wright's vision of evolution, the "shifting balance theory," when selection does then kick in, is a group selection not an individual selection. As he wrote in 1945: "selection between the genetic systems of local populations of a species ... has been perhaps the greatest creative factor of all in making possible selection of genetic systems as wholes in place of mere selection according to the net effects of alleles" (Wright 1945, 396). To make the pie complete, Wright himself added that a mechanistic explanation is at best just a surface explanation: "One may recognize that the only reality directly experienced is that of mind, including choice, that mechanism is merely a term for regular behavior, and that there can be no ultimate explanation in terms of mechanism – merely an analytic description"(Wright 1931, 154).

"The only reality directly experienced is that of mind." Schelling – who famously wrote "Nature is visible spirit, spirit is invisible nature" – would have hugged him.

Fruit Flies

Theodosius Dobzhansky's *Genetics and the Origin of Species* (first edition 1937, second 1941, third 1951) was arguably the most important – certainly the most influential – book on evolutionary theory in the twentieth century. This is even more true if you add in a subsequent dazzling series of papers on and around the topics of the book. What makes his work absolutely fascinating from our perspective is that, looking consecutively at the three editions, one sees a move – an evolution? – from the root metaphor of an organism to the root metaphor of the machine. In the first edition of his book, Dobzhansky rather set the scene by quoting in a disinterested fashion from two different perspectives, one (Fisher's) that makes selection all-important and that of two other contemporary writers that does not: "We do not believe that natural selection can be disregarded as a possible factor in evolution. Nevertheless, there is so little positive evidence in its favor ... that we have no right to assign to it the main causative role in evolution" (Dobzhansky 1937, 151, quoting Robson and Richards 1936, 316).

In this mode, Dobzhansky ran through some of the traditional evidence in favor of selection and in support of its importance. Melanism got short shrift. "Unfortunately the work on the industrial melanism has been restricted mainly to collecting the records of the happenings as they occur, and the causal analysis has lagged far behind, except for more or less gratuitous speculations" (160). Mimicry comes off little better: "Taken as a whole, an unprejudiced observer must, I think, conclude that an experimental foundation for the theory of protective resemblance is practically non-existent" (164). Dobzhansky was not entirely negative: "The adaptive value of the development of a longer pelage and a greater amount of wool in a cool climate is indeed obvious" (171). Dobzhansky was no Lamarckian, so it is selection at play here: "in a cold climate natural selection must *favor* the genotypes which, other conditions being equal, produce a warmer pelage" (emphasis in original). Overall, though, this is hardly the basis for a fully adequate theory of evolution.

Wright to the rescue! Dobzhansky, through a study of snails on oceanic islands, convinced himself that the variation between groups is nonadaptive: "The difficulty of proving that a given trait has not and never could have had an adaptive significance is admittedly great; nevertheless, the facts at hand are explicable, without stretching any logical point on the assumption that racial differentiation is due to mutations and random variations of the gene frequencies in isolated populations" (136). This set the scene for action by Wright's hypothesized scenario: "[A] colony that has reached a gradient leading to a new peak may climb it rapidly, increase in size, and either supplant the old species, or, more likely, form a new one that owes its allegiance to a new peak." Selection has a secondary, mopping-up role. "Natural selection will deal here not only with individuals of the same population (intra-group selection) but also, and perhaps to a great extent, with colonies as units (inter-group selection)" (190). In passing, remember that this latter process, intergroup selection, is something that Darwin explicitly eschewed. This, apart from the fact that Darwin, with his division-of-labor thinking, would be loath to think that there could be no adaptive differences between separate populations.

This issue, about adaptive influences, was something that led to a significant change in Dobzhansky's thinking. The third edition of *Genetics and the Origin of Species* (1951) is very different from the first edition, being far more selection/Darwin friendly. Dobzhansky's work on fruit flies, *Drosophila*, often in collaboration with Wright (who did the mathematics), had shown unambiguously that selection plays an active role in the life-cycles of small isolated populations. A 1943 paper on *Drosophila obscura* showed very significant genetic fluctuations that were clearly a function of ecological conditions at different times of the year. What might be behind all of this? A "plausible view is that Standard is favored while the populations are at their highest density levels in the year, and that Chiricahua is favored while the populations are dwindling toward their summer eclipse stage, presumably because of a relative food scarcity" (Figure 5.3).

Moving to the third edition of *Genetics and the Origin of Species*, when (to be fair) Ford and his group were in high gear, mimicry got a far friendlier treatment:

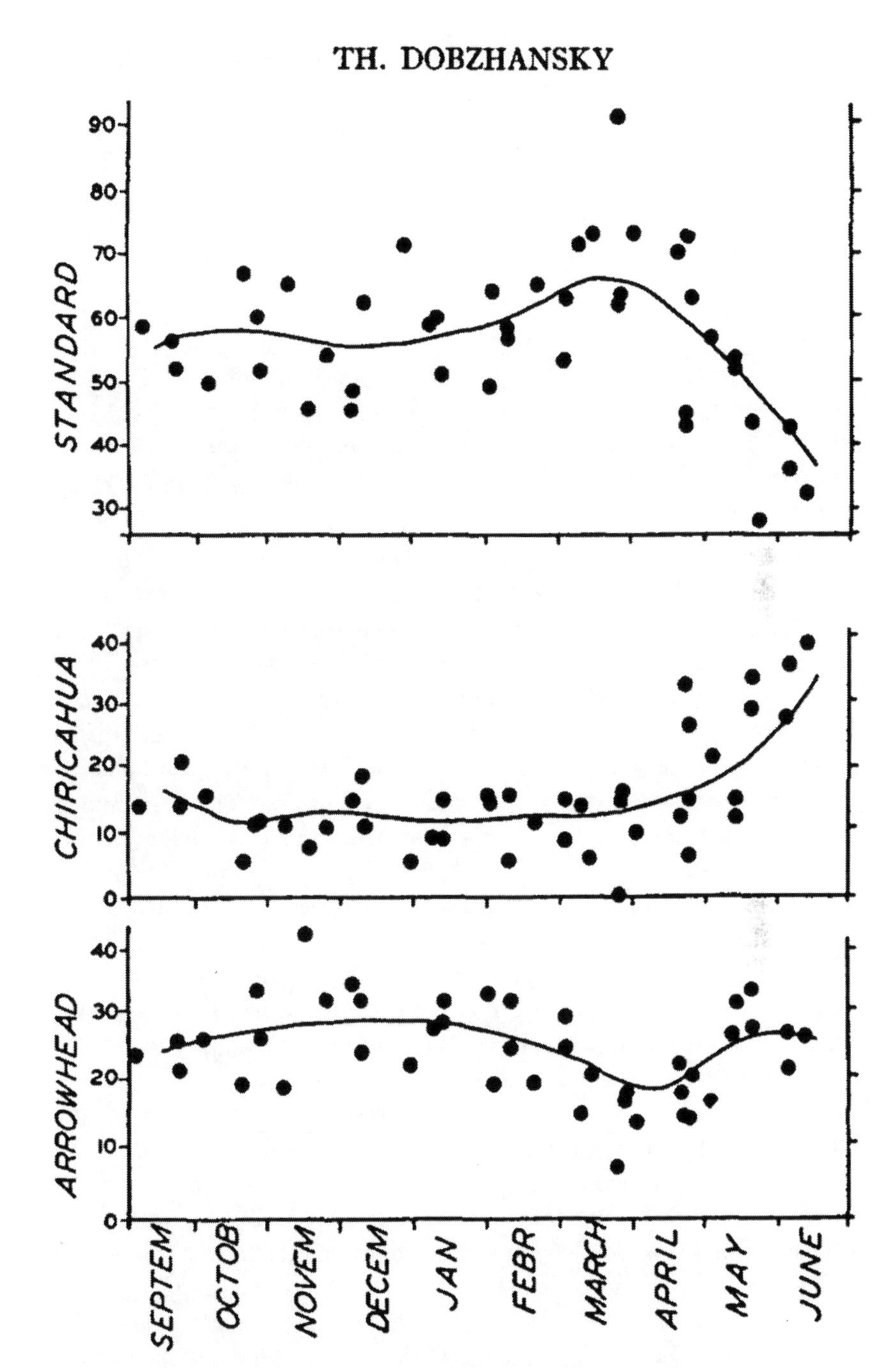

Figure 5.3 Temporal changes in chromosome types, suggesting that they are under the power of natural selection and not the result of drift.

> There can no longer be a reasonable doubt that many animals are camouflaged in their natural surroundings. It is of course, a different problem whether the camouflage has developed under the influence of natural selection, because of the protection from enemies which these properties confer on their carriers. Here one can proceed only by inference, with experiments pointing the way. Such experimental data as are available support the natural selection theory. (102)

Melanism likewise:

> The spread of melanic mutants was precluded before the advent of industrial developments owing to the destruction of such mutants by predators, since dark individuals are not protectively colored. This disability is removed in the industrial regions by the general darkening of the landscape. The superior viability of the melanics is able, then, to assert itself, and their rapid increase in frequency is the result. (132–33)

Things go the other way too. While Wright's theorizing about evolution is not dropped, the endorsement is nothing like as positive. Selection is given a much larger role: "It is not possible at present to reach definitive conclusions regarding the role played in genetic drift in evolutionary processes" (164).

By the time Dobzhansky had made his move away from drift towards a neo-Darwinian perspective – back to mechanism from neo-Spencerian organicism – he came face to face with the fact being uncovered by Ford and his associates, that selection does not just change genetic ratios but through such processes as balanced heterozygote superior fitness and selection for rareness maintains genetic variation in populations. This led him into a major row with Hermann J. Muller, one of Morgan's group, who not long before had won the Nobel Prize for his work on mutation, showing how X-rays could produce them artificially (Lewontin 1974). Muller argued that variation in populations is transitory, as selection removes new variants. Dobzhansky, thinking as a whole-animal evolutionist and not a molecular geneticist, was keen to challenge this. Why was this so important to Dobzhansky, other than the natural resentment towards a Nobel Prize winner, who automatically assumes that what he says necessarily is right? Dobzhansky was keenly aware that one of the main worries about natural selection is that, if the variations are random in the sense of nondirected, it seems implausible that if an organism has reason to evolve – a new predator for instance, or an environmental change such as an ice age – an appropriate variation will come along ready to be used. That is rather like having to write an essay on dictators and relying on the offerings of the Book of the Month Club (Ruse 1982). By the time anything vaguely useful came along, the course would be over and you would have failed. But now, thanks to selection's ability to maintain variation, it seems that you have a library at your disposal. If it doesn't have a book on Hitler, then perhaps one on Stalin, and if not Stalin then how about Mao? There is almost certainly going to be something. It is true that there is going to be no one favorable option, but that is as it may be. A new predator? Then perhaps there are a range of genes for coloration and one proves particularly useful for camouflage. Instead of the usual off-white, the move will be to a preponderance of dark grey organisms, who show up much less clearly against the vegetation. Or perhaps genes for vile taste, so the predators soon learn to avoid you. Then again, there might be genes that make organisms favor a nocturnal existence. And so the story goes, with perhaps genes that say "get the hell out of here fast." The whole point about natural selection is that it is relativistic. There is no one unique solution. No absolute progress. What works is what works.

And that is a good point to start celebrating the centennial of the *Origin*: a major conference at the University of Chicago with honorary degrees handed out like ice-creams (Smocovitis 1999).

6

Normal Science

Progress Resurgent

As explained in Chapter 3, Thomas Kuhn's influential notion of a "paradigm" is akin to but broader than the traditional notion of a scientific "theory" in that it includes the whole mindset that a theory generates – training and so forth. Most particularly, the sociological notion of having a foundation from which to work, to get on with the job. In this sense, one accepts the basic premises of the paradigm as given, not to be questioned, and then tries to solve problems generated by empirical research. Kuhn calls such problems "puzzles," because, as with a crossword, one knows there must be a solution, and failure to find it rests on the researcher rather than the paradigm itself. "Problems" are when it becomes clear that the paradigm is not adequate and one needs to think about finding a new paradigm. Paradigm changing is "revolutionary" science. Working within a paradigm is "normal" science.

With the general acceptance of the Darwinian theory of evolution through natural selection, backed by Mendelian genetics, now feeling free to use Kuhn's insights without necessarily buying into his idealism, evolutionary biologists now had their paradigm. One question, as we turn now to look at the paradigm in action: "normal science." The question of values. Grant that the Darwinian view of things is, *sensu* root metaphor, mechanistic. Adaptation is no longer a big problem. But what of the status of the highly pertinent notion of progress? In the case of Sewall Wright, who was clearly thinking in such terms, the value comes with the root metaphor. In the case of Darwin, we have seen that he himself had difficulty purging his science. And moving forward in time, we shall see other important evolutionists facing this issue.

To unpack this issue, let us make a philosophical distinction between "epistemic values" about science and "nonepistemic values" in science (McMullin

1983). The former are rules, such as predictive fertility and falsifiability and consilience of inductions and, perhaps, simplicity, aimed at uncovering the objective truth about the real world. The latter involve topics such as racism, feminism, and antisemitism, aimed at creating a product that confirms our value systems, often prejudices. If we are not being, consciously or unconsciously, philosophical, when speaking of thinking of nonepistemic values we usually refer to them simply as "values," without qualification. Typically, speaking now of work (such as Darwin's) governed by the machine root metaphor, we have a body of knowledge claims starting off as a pseudoscience governed by nonepistemic values, in our case progress. People certainly said that of *Vestiges*, and we have seen reason to think that it applies to Erasmus Darwin's thinking. Increasingly, epistemic values take over, and push out the nonepistemic values. One starts with alchemy and ends with chemistry. One starts with mesmerism and ends with psychology.

The history of progress as embedded in evolutionary thinking seems to fit this pattern (Ruse 1996). It is nonepistemic. One is making a value judgment about the superior worth of human beings. Monad to man. Note that it can be contested as to whether something is epistemic or nonepistemic. Someone like the mechanist Darwin thinks that the concept of progress is nonepistemic. Organicists are a little trickier to categorize. Lamarck would argue that he escapes this censure because he was not a mechanist, and he does not regard progress as a nonepistemic value. Spencer likewise. Which explains why a mechanist would probably judge all organicist thinking pseudoscience! Organicists might argue that progress is not nonepistemic, or they might argue (one suspects more likely) that it is simply nonsense to think that science ever escapes nonepistemic values.

Historically, we have seen that, for the first hundred years of the idea of organic evolution, most practicing (professional) scientists, being Newtonian, considered the idea to refer to somewhat of a pseudoscience, because it was governed by the nonepistemic value of progress. Note, however, that the story is a little more complex than simply downgrading/dismissing evolutionary thinking as not real science. The point is that the critics such as Sedgwick and Whewell disliked the particular nonepistemic value – progress. So such thinking was cast aside. The critics were not against nonepistemic values as such. Their nonepistemic value was Providence. But the result of endorsing this was not a rival would-be science. Rather, the topic of origins was taken out of science and placed in religion! So the alternatives were evolution – pseudoscience – or divine creation – not science at all. Then came Darwin, and at least in the science he produced – we shall later consider other areas such as philosophy, religion, and literature – progress was booted out. Natural selection precludes inevitable progress. Human brains are not the inevitable end point. In the memorable words of the paleontologist Jack Sepkoski: "I see intelligence as just one of a variety of adaptations among tetrapods for survival. Running fast in a herd while being as dumb as shit, I think, is a very good adaptation for survival" (Ruse 1996, 486). Evolution was on its way to being a real science.

Although note here, things were not entirely smooth. Darwin did not want progress in his science, and those using natural selection – people such as Bates – simply were not interested in progress. If the bark is sooty, dark coloration is what is needed. If the bark is clean, light coloration is needed. It is all relative. Against this, those working on fossils and the like were clearly drawn towards progress. Yet here is the interesting thing. Generally, these people, although they believed in evolutionary progress, did not consider the science as such to be real professional science. Huxley barely lectured in class on it and Hyatt kept it out of his textbook. Evolution was more for the popular audiences – at working men's clubs and the like. In other words, it was more a kind of popular science, such as one finds in museums and elsewhere. Places where values are acceptable. After all, a primary function of museums is indoctrinating children in the right values – all humans are of the same intrinsic worth, that sort of thing. Understood for what it is, popular science can be very important and of great moral worth.

We move into the twentieth century and up to the centenary of the *Origin*, 1959. Now we have an evolutionary theory based on natural selection and no progress is presupposed or promoted. Although, as it turns out, even here the story is a little more complicated. Progress may no longer have been the norm in *professional* evolutionary biology. One should not mistakenly think that the scientists were no longer progressionists. To the contrary, most were ardent enthusiasts for progress! It is just that they thought that by embracing so obvious a nonepistemic value in their science, they would violate the (nonepistemic) demand that professional science has no nonepistemic values driving it. A kind of meta-nonepistemic value prohibiting nonepistemic values in science. Otherwise, one faces the end of fancy professorships, the end of grants, the end of fellowships of much venerated societies. As we have seen, sometimes the nonepistemic progress just slips in. Darwin eschewed progress but ended the *Origin* by declaring: "from the war of nature, from famine and death, the most exalted object which we are capable of conceiving, namely, the production of the higher animals, directly follows" (Darwin 1859, 490). Sometimes it is deliberate. Talk of progress is allowable if one is clearly talking at the *popular* level, as in museums – as we have just seen – or in books for the general reader. G. G. Simpson published *Tempo and Mode in Evolution* in 1944, his paleontological contribution to the synthetic theory. No hint of progress. Then *The Meaning of Evolution* in 1949, a book for a popular audience. Progress, openly discussed and embraced. Finally, *Major Features of Evolution* in 1953. An update of *Tempo and Mode in Evolution*. No progress.

Simpson was far from atypical. It is interesting to speculate as to why progress was/is so popular. Possibly because science is the one area of human effort where there does seem to be progress. The heliocentric theory of the universe is better than the geocentric theory, sort of thing. One doubts that theater directors think this way. A production of *Richard III* in Victorian times, at the height of the British Empire, would probably be very different from a production in the late 1930s, when the Third Reich was up and running and threatening worldwide war. It is not a question of which production is better; they are tuned in to different ages. The Victorian production, if staged at the time of the Indian Mutiny, might have Richard looking like a maharaja. The 1930s production might have Richard in a German uniform with a swastika. Science is different. The geocentric theory is not tuned in to the age of ancient science and the heliocentric to the age of modern science. One is wrong and one is right. Progress!

Normal Science?

Picking up now on our general topic, does one see evidence of normal science? One certainly does. Take the problems that arise in trying to understand social behavior (Burkhardt 2005). From the first, social behavior was considered as important as the fossil record or the facts of geographical distribution. Darwin had insightful things to say, especially in the context of his commitment to an individual-selection perspective. However, as one might have expected, for the next hundred years, social behavior was always somewhat of a poor relative considered next to other areas such as paleontology and biogeography. Social behavior can be so difficult to observe and quantify, a problem multiplied by the fact that animals in captivity so often do not do what they do in the wild, or they do what they do not do in the wild. What in the wild are separate noncrossbreeding species copulate happily and successfully in the zoo, or conversely while fully fertile in nature refuse absolutely to breed in the zoo, despite multiple enticements.

It was not until the 1930s that a number of continental biologists, given the name "ethologists," started serious observation of behavior, particularly social behavior. Noteworthy was the Austrian animal biologist Konrad Lorenz, who became famous for his studies of imprinting of geese – how they become attached to and follow the movements and directions of their caregivers. By the time of the Darwinian anniversary, Lorenz and others, notably the Dutch zoologist Niko Tinbergen, were well-known and respected biologists (Lorenz 1966; Tinbergen 1968). One noteworthy point is that, despite thinking of themselves as Darwinians, group selection was considered a viable and important force of evolutionary change. Although this did not apply to all – Tinbergen, for example – in the 1930s, Lorenz was a Nazi: "I can say that my whole scientific work is devoted to the ideas of the National Socialists" (World Jewish Congress 2015). Holism was a major philosophical underpinning of the Third Reich's concept of a state (Harrington 1996). After the war, vile ideologies now discarded (or not mentioned), Lorenz and fellow enthusiasts for group selection were supported strongly by a growing number of Anglophone biologists, most notably V. C. Wynne-Edwards, author of *Animal Dispersion in Relation to Social Behaviour* (1962).

This assumption of group selection sparked a strong reaction from a growing group of neo-Darwinian evolutionists. Most significantly, in the early 1960s, the then graduate student William Hamilton followed Darwin in trying to explain the social behavior of the hymenoptera (ants, bees, and wasps) in terms of natural selection (Hamilton 1964). What could be more undeniably an effect of group selection than the fact that the female workers are sterile and spend their lives laboring for the good of the nest? Hamilton pointed out, however, that the hymenoptera have an atypical mating system. They show "haplodiploidy." Females have both mothers and fathers – they are the result of the union of equal shares of DNA from female and male (and hence are diploid, with two half-sets of chromosomes) – whereas males have only mothers – they are born of unfertilized eggs, and hence have only the DNA of their mothers (and consequently are haploid, having only one half-set of chromosomes). This means that, unlike normal reproductive systems where females are as closely related to daughters as to sisters (50 percent), in hymenoptera the females are more closely related to sisters than to daughters (75 percent to 50 percent). Hence, from a selective viewpoint – where what counts is getting copies of one's DNA into the next generation – females maximize their own reproductive chances by raising fertile sisters rather than fertile daughters. A triumph of individual selection, because the infertility of the female workers is to their own adaptive advantage!

Sociobiology

Thanks to Hamilton's work and a host of subsequent studies thinking and working in the mode he initiated – George Williams's *Adaptation and Natural Selection* (1966) was rightly lauded as an important contribution – as we move into the next decade, the 1970s, Darwinian evolutionists had reason to feel confident, if not smug. Work showing the power of thinking in terms of the survival and reproduction of individual genes, where family connections are all-important – newly christened "kin selection" – abounded. The whole field was glamorized and brought to public attention by the publication of *The Selfish Gene* (1976), a metaphor-laden book explaining the work and its opportunities, by the young Oxford biologist Richard Dawkins. Overshadowing all was a massive overview of the field by the Harvard ant biologist Edward O. Wilson. *Sociobiology: The New Synthesis* (1975) gave a detailed review of the theory and then gave an account of the evolution of social behavior, from the (presumed) lowest forms of animals up to the summit of the process, *Homo sapiens*. Our behavior, our social structures, were all explained in terms of what Dawkins was to label "selfish genes." Expectedly, the nuclear family is all-important:

> The populace of an American industrial city, no less than a band of hunter-gatherers in the Australian desert is organized around this unit. In both cases, the family moves between regional communities, maintaining complex ties with primary kin by means of visits (or telephone calls and letters) and the exchange of gifts. During the day the women and children remain in the residential area while the men forage for game or its symbolic equivalent in the form of barter and money. (Wilson 1975, 553)

And so on: "Sexual bonds are carefully contracted in observance with tribal customs and are intended to be permanent. Polygamy, either covert or explicitly sanctioned by custom, is practiced predominantly by the males" (553–54).

A somewhat conservative vision – emphasis on different male and female roles, for instance – but nothing execrably offensive, belittling black people or Jews or that sort of thing. Initial reaction was impressed and favorable. British geneticist Conrad Waddington wrote a positive review in the *New York Review of Books*. Then all flared up into a huge controversy, with harsh divisions akin to those separating American political parties today. Taking the leading roles were biologists Stephen Jay Gould and Richard Lewontin (Reiss and Ruse 2023). Wilson was accused of every possible sin, from methodological and evidential weakness to outright racism – promoting the status quo of the haves against the have nots. What made everything particularly bitter and personal was that Gould and Lewontin were fellow members of the Harvard biology department, along with Wilson. The motives for the attack were mixed. Gould and Lewontin were both Jews, and Gould particularly felt that Wilson's discussion opened the door to antisemitism. Backing this charge, he wrote a book: *The Mismeasure of Man* (1981). Lewontin's major motivation was political philosophy (Lewontin 1976). His most recent previous appointment had been at Chicago, where he was radicalized by fellow biologists – most importantly by the ecologist Richard Levins, who also moved to Harvard. He was especially spurred by the ongoing war in Vietnam and became a fervent Marxist, a philosophy shared by Gould, if never in quite such an ideological way.

Showing that deeper and more venerable motives were at work, in reaction to what they saw as the unwonted, mechanism-based approach of Wilson, in a much quoted paper – "The spandrels of San Marco and the Panglossian paradigm: a critique of the adaptationist program" – they declared their full-blooded acceptance of the organismic approach (Gould and Lewontin 1979). For Gould particularly, this was not a particularly innovative move. In earlier writings, he made heavy use of a close relative of Sewall Wright's genetic drift, the so-called "founder principle," the claim that small, isolated populations (say, immigrants to an island) are going to be statistically different from their parent populations, because the variation in the parent populations means that the small populations will almost certainly have more of some alleles and fewer of others (Eldredge and Gould 1972; Gould and Eldredge 1977). As with drift, there is a random factor here – a flock of birds blown out to sea in a hurricane will, by chance, be statistically different from their parent population. There is also selection, both in creating the variation in the parent population and, almost certainly, in the newly isolated flock, drawing on a depleted gene library in a new environment, which is going to have hitherto-unfaced adaptive demands: new predators and so forth. Because of the intensity of the selective forces, one expects rapid change – change too fast to be recorded in the fossil record.

By the end of the decade, Gould (1980) was arguing more vigorously in a nonselection-friendly mode. At least part of the motivation for the shift was that Gould was ardently promoting the essential role of paleontology in the Darwinian synthesis. Although the general public usually thinks first of the fossil record when they think of evolution – remember *Evolution: The Fossils Say No!* (Gish 1973) – Gould felt, with some reason, that geneticists and others working on causes tend to belittle the physical evidence of the past. The Nobel-Prize-winning molecular biologist Peter Medawar referred to a prominent paleontologist (Teilhard de Chardin) as a "naturalist" practicing "a comparatively humble and unexacting kind of science"(Medawar 1961, 101). The message of the nineteenth century was that the fossil record is to be explained rather than looked up as something able to do the explaining. Gould was determined to up the status of his discipline – to claim a seat at the "high table" (referring to the privileged place of faculty in the dining halls of Oxbridge colleges) – and to this end he was speculating in non-Darwinian jumps, saltations, as the essential force behind his vision of evolution, "punctuated equilibrium."

Whether or not any of this was going to appeal to the geneticist Lewontin – we shall shortly see reason why it would not appeal – this was the height of

Figure 6.1 Spandrels of San Marco.

the sociobiology controversy with Lewontin uniting with Gould in the attack (Allen et al. 1975, 1976). Downplaying one's own field and achievements was small payment to take the battle to the enemy. To this end, belittling the credentials of Wilson's Darwinian science, Gould and Lewontin argued that natural selection is not the main force for change. To the contrary:

> We criticize this approach and attempt to reassert a competing notion (long popular in continental Europe) that organisms must be analysed as integrated wholes, with Baupläne [a modern equivalent of Owen's archetypes] so constrained by phyletic heritage, pathways of development and general architecture that the constraints themselves become more interesting and more important in delimiting pathways of change than the selective force that may mediate change when it occurs. (Gould and Lewontin 1979, 147)

They focused on the triangular areas at the tops of columns in St. Mark's Church in Venice (Figure 6.1). These are covered with exceptional mosaics; but, the point of the Gould–Lewontin argument is that the spandrels were not made – selected – for this purpose. They are rather the side effects of the architecture needed to keep the roof in place. The spandrels seem to be designed for what they do. "Yet evolutionary biologists, in their tendency to focus exclusively on immediate adaptation to local conditions, do tend to ignore architectural constraints and perform just such an inversion of explanation" (Gould and Lewontin 1979, 149). This opens the way for a good jab at – a strong, hopefully fatal thrust at – an argument by E. O. Wilson, that Aztec human sacrifice is an adaptation to provide fresh meat. "We strongly suspect that Aztec cannibalism was an 'adaptation' much like evangelists and rivers in spandrels, or ornamented bosses in ceiling spaces: a secondary epiphenomenon representing a fruitful use of available parts, not a cause of the entire system" (584).

From here Gould and Lewontin launched into a general critique of Darwinism. "We wish to question a deeply engrained habit of thinking among students of evolution. We call it the adaptationist programme, or the Panglossian paradigm" – after Dr. Pangloss of Voltaire's *Candide*, who saw everything as being part of the "best of all possible worlds." As we have just seen, as an alternative they turned to a continental position focusing on *Baupläne*, the underlying archetypes structuring organisms. Natural selection cannot explain them. We need another approach: one that "holds instead that the basic body plans of organisms are so integrated and so replete with constraints upon adaptation ... that conventional styles of selective arguments can explain little of interest about them." Condescendingly: "It does not deny that change, when it occurs, may be mediated by natural selection, but it holds that constraints restrict possible paths and modes of change so strongly that the constraints themselves become much the most interesting aspect of evolution" (594).

As we prepare to move on, it should be noted that history shows the paradoxical – better, farcical – nature of this clash. The Spandrels paper was written from a holistic, organismic perspective. Yet, as a practicing scientist, Richard Lewontin was a totally committed pro-Darwinian, mechanistic reductionist! With the path-breaking discovery in 1953 of the double helix, by James Watson and Francis Crick (1953), biologists had an entry into the molecular workings of heredity. Not an overthrow of the Mendelian genetics so painstakingly unraveled in the first half of the century, but putting it all on a molecular basis, so one could more accurately and fully unravel the mysteries of heredity. Lewontin, whose graduate supervisor was Theodosius Dobzhansky, learnt at firsthand of the clash between Dobzhansky and Muller over the possibility of variation being a feature of all natural populations. In the 1960s, he was one who moved right into the molecular field, developing the powerful tool of gel electrophoresis, which was able to show beyond question that such variation as Dobzhansky needed was in fact always present (Lewontin 1974). No need to wait on something good appearing. A triumph of normal science, where molecular biology was shown to be a powerful tool for the conventional whole-organism evolutionist – Darwinian evolutionist – solving old problems rather than threatening a wholesale demolition or downgrading of what had come before.

The inference is that Lewontin allowed his name to be attached to the Spandrels paper less because he believed the science (or philosophy behind the science) and more because this was a weapon to attack Wilson. Gould – who, from the style of the paper, was almost certainly its lead author – was innocent of such hypocrisy. As a paleontologist he had no commitment to Darwin, and punctuated equilibrium seems more a child of the Spencer/Sewall Wright school of thought. We learnt that in the case of Bateson's attitude to his teacher, Weldon, Oedipal forces were plausibly at work. The same forces may well have been operative here. Lewontin was given to belittling Dobzhansky, virtually sneering at him because of his supposedly limited skills: "a theoretician without tools" (Lewontin 1995). "At times, this worry – and Lewontin's crowing about his superior command of some branches of mathematics, especially the statistical analysis in which he had an MA – left Dobzhansky bewildered and hurt" (Depew 2023). Unsurprisingly: "In 1963, he told his former student, 'I am so used to your disapproval that it no longer hurts me as much as it used to' (Dobzhansky to Lewontin, May 7, 1963)." Spandrels, published a year or two after Dobzhansky's death, fits nicely into this picture.

The paradox or farce is even more extended than all of this. Far from being a dyed-in-the-wool mechanist, Wilson always had organicist leanings (Gibson 2013; Reiss and Ruse 2023). On the wall of his laboratory there was a picture of Herbert Spencer, larger, more prominently placed than the picture of Charles Darwin. "Great man, Mike! Great man!" was the response to this incredulous author. Like Sewall Wright, Wilson was influenced by the Spencerian enthusiasms of the early twentieth-century Harvard biologists, in his case by his intellectual grandfather, the ant specialist William Morton Wheeler, the supervisor of his supervisor. Wheeler was enthusiastic about analogies between human and ant societies, going on to speak of ant societies as being akin to individual organisms. To be candid, conversely, Wheeler was not overly enthusiastic about natural selection (Wheeler 1923, 1939; Evans and Evans 1970). Like Spencer, he emphasized cooperation as an important aspect of organic life as opposed to struggle. Wheeler was much interested in the implications of all this for our human society, and like Spencer he had no doubt that humans were the triumphant end of the evolutionary progress. Virtually all of this comes through unchanged in Wilson's vision of evolution. Right at the heart of his science, he was nigh fanatical about progress with humans at the top. He tells us that of all animals, "Four groups occupy pinnacles high above the others: the colonial invertebrates, the social insects, the nonhuman mammals, and man" (Wilson 1975, 379). He continues: "Human beings remain essentially vertebrate in their social structure. But they have carried it to a level of complexity so high as to constitute a distinct, fourth pinnacle of social evolution" (380). He concludes by speaking of humans as having "unique qualities of their own" (380). He then launched at length into showing us how humans have crossed over and mounted the "fourth pinnacle" (382) – the "culminating mystery of all biology" (382). All this, as Wilson made clear in subsequent writings, is very much part of the general picture:

> The overall average across the history of life has moved from the simple and few to the more complex and numerous. During the past billion years, animals as a whole evolved upward in body size, feeding and defensive techniques, brain and behavioral complexity, social organization, and precision of environmental control – in each case farther from the nonliving state than their simpler antecedents did. (Wilson 1992, 187)

And if this is not enough to convince you of Wilson's organicist credentials, there was his open embrace of group selection: "Four decades of research since the 1960s have provided ample empirical evidence for group selection, in addition to its theoretical plausibility as a significant evolutionary force" (Wilson and Wilson 2007, 334). As we might expect, human evolution is part of the argument: "Group selection is an important force in human evolution in part because cultural processes have a way of creating phenotypic variation among groups, even when they are composed of large numbers of unrelated individuals." Genetic mutation is slow working: "If a new behavior arises by a cultural mutation, it can quickly become the most common behavior within the group and provide the decisive edge in between-group competition" (343).

Almost amusingly, 137 (!) of Wilson's fellow evolutionists penned a critical letter to *Nature* (Abbott et al. 2011). Against the just-quoted article, "their arguments are based upon a misunderstanding of evolutionary theory and a misrepresentation of the empirical literature." Need more be said? Given the inwardly conflicting positions of both Wilson and Lewontin, one is tempted to give a sociobiological explanation – rival Harvard alpha males striving for dominance – and leave matters at that.

Towards the Present

Over the past half-century and more since the Darwin centenary, it is surely true – Harvard tantrums notwithstanding – evolution through natural selection has settled into the dominant paradigm, within which biologists can do normal science, and that indeed is what they do. Today, natural selection is accepted easily and readily by working evolutionists. To take but one example, the deservedly highly praised work on the evolution of Darwin's finches in the Galapagos, of the husband and wife team Peter and Rosemary Grant (Grant and Grant 2003). In one celebrated study, they tied changes in the food supply of one species of finch to a severe El Niño event. Large seeds became scarce and so there was a knock-on effect of natural selection bringing on small beak sizes. The moral? "The evolutionary change documented in the population of *G. Forli's* on Daphne serves as a model for what has presumably happened countless times on a larger timescale in the past: small evolutionary changes in quantitative traits caused by natural selection under changing environmental conditions" (Grant and Grant 1993, 116).

To take a second example, from a paper about microevolutionary changes in the finches: "The best scope for a predictive theory of evolution lies in the area of genetics, because the mathematical machinery has been developed for the precise prediction of evolutionary change caused by selection" (Grant and Grant 1995, 241). And a third example, from an abstract about species diversity: "Darwin's finches on the Galápagos Islands are particularly suitable for asking evolutionary questions about adaptation and the multiplication of species: how these processes happen and how to interpret them" (Grant and Grant 2003, 965). And how does it all happen? "Key factors in their evolutionary diversification are environmental change, natural selection, and cultural evolution. A long-term study of finch populations on the island of Daphne Major has revealed that evolution occurs by natural selection when the finches' food supply changes during droughts" (Grant and Grant 2003, 965) (Figure 6.2).

The Grants' Galapagos studies are about as typical an instance of normal science as it is possible to imagine. They picked up the Darwinism of 1959, and using its theory showed its power. Theirs was an understanding of bird behavior that was unprecedented. Is this the end of the story? Not at all. One topic thus far unmentioned is speciation. In the *Origin*, Darwin stressed that varieties seem to follow an unbroken line as they move towards being species, but he has little discussion of the defining nature of a species or how they are formed. This has been a topic of intense interest to today's evolutionists. Species are usually defined in terms of reproductive isolation. Prominent questions are whether species are formed simply by chance, as it were – groups are physically isolated and reproductive barriers are a function of evolving in different directions because of different selective needs ("allopatric" speciation) – or if the different paths reflect the need for different adaptations in a continuous group, perhaps with different ecological niches, where species barriers are an adaptive need ("sympatric" speciation). Discussions of these topics often give one a sense of what the religious mean when they talk of "eternity." A good introduction to the topic is by Jerry A. Coyne and H. Allen Orr (2004).

This all said, is it the case that, despite mavericks such as Wilson, the machine metaphor rules supreme? No indeed! For all the successes of work such as that of the Grants, philosophy often trumps fact and theory.

Figure 6.2 Four of the species of finch (known now as "Darwin's finches") that Darwin saw on the Galapagos. The different beaks show different adaptive strategies. For instance, the large beak on the bird top left is for cracking nuts and seeds; the fine beak on the bird bottom right is for catching insects. (From Charles Darwin, *Journal of Researches*, 1845, 379.)

Organicist-based criticisms of the conventional Darwinian theory abound. One might think that, in the light of Lewontin's successful foray into molecular biology to confirm Dobzhansky's beliefs in intraspecies variation, here is the kind of work that will always promote a mechanistic approach to evolution. And, given the high status of this ultra-reductionistic approach to organisms, there will be a kind of slop-over effect to all studies of biological evolution. Not so! Almost paradoxically, it is molecular biology that gives most hope to organicists (Reiss and Ruse 2023).

Start with the fact that when the synthetic theory was put together, in the 1930s, development was a bit of a black box. Other than Gavin de Beer, most embryologists were not that much interested in evolution and most evolutionists were ignorant of embryology. No one was about to deny the potential importance of organic development, but first things first, and so there was a tendency to treat organisms somewhat like sausage machines.

Genes – genotypes – in; organisms – phenotypes – out. Understanding what happens in the sausage machine itself is a task left until later. And with the advent of molecular biology, "later" had arrived (Arthur 2021). Molecular biologists were already tracing the paths of information on DNA molecules to information on RNA molecules, and then from RNA molecules to amino acids, with the information being used to order them and thus make proteins of different kinds. One was well on the way now, as biologists traced the paths from proteins to full-blown organisms, phenotypes. This new subfield of evolutionary development, "evo-devo," seemed a textbook example of mechanistic science in action. "Evo-devo provides a causal-mechanistic understanding of evolution by using comparative and experimental biology to identify the developmental principles that underlie phenotypic differences within and between populations, species and higher taxa" (Laland et al. 2015, 3).

Nevertheless, evo-devo soon attracted those with organicist yearnings, and we were on our way to an impassioned attack on pure mechanism as evinced by natural selection. The possibility was opened that significant biological change – adaptation – comes not from without – the sorting and successes and failures of different phenotypes – but from within – the individual organisms in some way control their development into adaptively successful organisms. From the primitive cell to the fully functioning human being. The key is internal development rather than being guided exclusively by external forces. "While much evo-devo research is compatible with standard assumptions in evolutionary biology, some findings have generated debate. Of particular interest is the observation that phenotypic variation can be biased by the processes of development, with some forms more probable than others" (Laland et al. 2015, 3). Numbers of digits and vertebrae are instanced as supportive of this claim.

Expectedly, conventional evolutionists – labeled Standard Evolution Theory (SET) supporters – respond to this charge, claiming that the true, complete answers support their case rather than that of their organicist critics – Extended Evolution Synthesis (EES) supporters. Take, for example, cave fish. The Mexican tetra, *Astyanax mexicanus*, has invaded caves, and many forms, independently, have started to lose their sight. The eyes start off normally, but then they stop developing, and in some cases this leads to total blindness. Adaptive reasons for this are easy to produce. Perhaps in the cave environment, eyes are of little use. In fact, they might be dangerous because the detritus in the cave could lead to irritations and infections; or other such reasons. However, blind cave fish have been seized on by EES enthusiasts as evidence of environmentally induced change that then gets incorporated in the gene line. Directed variation, meaning that you get a variation of use straight off without selection, produced by the organism itself. Lamarckism!

Eva Jablonka, a well-known enthusiast for the EES, "thinks that heritable epigenetic changes alone could explain the loss of eyes. What is more, she even thinks it possible that the epigenetic changes were somehow triggered by the cave environment in the first place. That would be a form of Lamarckian evolution: the idea that characteristics acquired during an individual's lifetime can be passed on to descendants" (Le Page 2017). SET supporters are not impressed. "'This is a most interesting paper,' says evolutionary biologist Douglas Futuyma of Stony Brook University in New York. But he doesn't think it poses any challenge to standard evolutionary theory as the epigenetic change is itself most likely a result of a genetic change" (Le Page 2017).

How do we disentangle the opposing positions? Apparently, what is causing the eye loss is a phenomenon known as "methylation." This occurs when a set of molecules – a methyl group – attaches itself to other molecules, preventing or altering their function, such as a DNA molecule no longer being able to give out information – to take our case – on how to produce eyes. What excites someone such as Jablonka is that these methyl groups are activated by environmental conditions – being in a cave with no light in this case – and then apparently go backwards to the DNA molecule where they become part of it. And, no denying, this can go on indefinitely. Inheritance of acquired characteristics!

Hardly, say SET supporters. A student of Lewontin, Jerry Coyne, writes of what is going on here:

> The thing is, this is not a violation of evolutionary theory in any sense. The commands for methylation under certain conditions, and the results of a gene being methylated, have evolved by changes in regular DNA sequences that control the methylation of other DNA sequences. It's just an evolved way to regulate genes. What has likely happened is that these cave fish, then, is that other, "regulatory" genes have changed in the cave forms that provide instructions sort of like this: "hey, gene Y: at a certain point, you attach methyl groups to other genes for eye formation, shutting down those genes and causing the eyes to degenerate." That could easily evolve by conventional natural selection. (Coyne 2017)

There are pithier comments in Coyne (2019).

We'll pause here to put things in context. There are going to be huge adaptive advantages in organisms being able to regulate their development according to environmental conditions. For instance, if an organism grows in an environment with limited food sources, there will be big adaptive advantages to growing up small with a correspondingly small appetite (Veenendaal et al. 2013). Conversely, in an environment with lots of food, growing up big with a large appetite may protect you from predators or – as with high school and football players – get you a pretty girlfriend. From the perspective of natural selection, much more efficient for doing this sort of thing, rather than having to produce two sets of genes in two sets of organisms, where one set succeeds and the other set does not, is to have ways to regulate growth along the way, in one set of organisms. This is where methylation comes into play. It enables the organism to switch off genes and so forth, having ultimate phenotypic consequences. You don't need to start everything from scratch. But where do the instructions from methylation come from? Ultimately, the DNA. There are genes that tell the methyl groups to spring into action or not. And these controlling (regulating) genes are there because of natural selection. Unless they confer advantage – for instance, giving the organism the ability to vary its size and appetite adaptively – they are not going to be around for long.

Leave it at this. The reader can see how the arguments are framed and presented: how an EES enthusiast is going to make the case and how the SET supporter will respond. The problem is that nature provides new challenges – let us say a new predator or a change in climate. How is the organism to respond? For the EES supporter, in some sense, you need directed mutations. Changing color to avoid the predator. Growing a furry coat to protect against the cold. Not so fast, says the SET supporter. Organisms, thanks to such phenomena as selection for rareness, always have a range of (genetically caused) variations to call on. No appropriate camouflage variations? Then how about a variation to move to a nocturnal existence? Or to become so disgusting in taste that the predator seeks elsewhere? Stay with mechanism. No need to move to organicism with its flavors of teleology. You have an end? Let nature supply it.

Finished!

And now, talking of ends, let us finish our overview of the science since the *Origin*. There are today enthusiasts for an organismic approach – enthusiasts, with very loud voices, promoting their beliefs. Generally, however, the fully mechanistic Darwinian theory of evolution through natural selection is the accepted norm – paradigm – for today's evolutionary biologists. Rosemary and Peter Grant are the Platonic Forms of today's evolutionists, both in their being a template for what it means to be a professional evolutionist and for the excellence of their work. They are both fellows of the Royal Society in London, as well as of the Royal Society of Canada. There is even a Pulitzer Prize-winning popular account of their work, telling us what it all means (Weiner 1994)! Which seems a good point on which to turn to more humanistic issues.

7

Philosophy

WHEN STRUCTURING THE ARGUMENT OF THE *ORIGIN*, DARWIN drew heavily on the philosophers – the empiricist John F. W. Herschel and the rationalist William Whewell (Ruse 1975b). Not much thanks he got from them after he published. We have seen that Herschel referred to natural selection as "the law of higgledy piggledy." Whewell, as Master of Trinity, refused to allow a copy on the shelves of the college library. Others, if not quite so negative, showed a noteworthy lack of enthusiasm. The response of John Stuart Mill was the epitome of damp praise: "Mr. Darwin has never pretended that his doctrine was proved. He was not bound by the rules of Induction, but by those of Hypothesis. And these last have seldom been more completely fulfilled. He has opened a path of inquiry full of promise, the results of which none can foresee" (Mill 1862, 2, 18n).

Humankind

Philosophy is about human beings, what they know – "epistemology" – and what they should do – "ethics." After the *Origin* appeared, Thomas Henry Huxley at once brought up the human question. He and Owen got into a bitter dispute over whether humans had a part of the brain known as the "hippocampus minor" and gorillas did not, Huxley arguing that both had the respective part (hence pro-evolution) and Owen arguing that it was unique to humans (and hence anti-Darwinian). Bitter, although not beyond joking. In his children's story, *The Water-Babies*, its author the Reverend Charles Kingsley has Professor Ptthmllnsprts tell us that:

> Nothing is to be depended on but the great hippopotamus test. If you have a hippopotamus major in your brain, you are no ape, though you had four hands, no feet, and were more apish than the apes of all aperies. But if a hippopotamus major is ever discovered in one single ape's brain, nothing

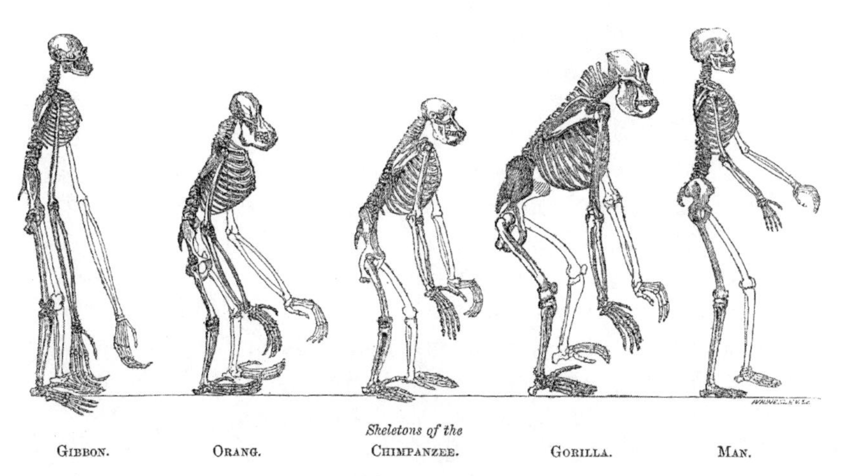

Figure 7.1 Frontispiece of T. H. Huxley's *Man's Place in Nature* (1863). Note that the gibbon has been drawn at twice its size to make it seem more that it is part of a series.

> will save your great-great-great-great-great-great-great-great-great-great-great-greater-greatest-grandmother from having been an ape too. (Kingsley 1863, 191–92)

In respects, Huxley was somewhat reserved on the topic of human ancestors. Neanderthal fossils were now being unearthed, but Huxley was hesitant about reading too much into them. In his *Man's Place in Nature* (1863), he wrote that in no sense "can the Neanderthal bones be regarded as the remains of a human being intermediate between Men and Ape" (206). However, when it came to the basics, as the frontispiece shows clearly, there was no ambiguity about the fact that humans are descended from the apes. "So far as cerebral structure goes, therefore, it is clear that Man differs less from the Chimpanzee or the Orang, than these do even from the Monkeys, and that the difference between the brains of the Chimpanzee and of Man is almost insignificant, when compared with that between the Chimpanzee brain and that of a Lemur" (140) (Figure 7.1).

Ask about Charles Darwin himself: "Light will be thrown on the origin of man and his history" (Darwin 1859, 488). One senses that, having finished the *Origin*, Darwin had little interest in following up his brief comment. His next major project was orchids, a topic he explored vigorously for a couple of years (Darwin 1862). Not exactly what one might have expected from the author of the most important book in the history of the life sciences. Darwin's noninvolvement in the evolution of our species ended with the apostasy of Wallace. By the mid 1860s, the codiscoverer of natural selection was showing troubling signs of entanglement with pseudoscience. He was increasingly vocally vehement that the only plausible explanation for the evolution of humankind was the action of spirit forces.

Appalled, Darwin set about giving a naturalistic explanation involving selection. *The Descent of Man and Selection in Relation to Sex* appeared in 1871. While I doubt anyone was thinking explicitly in these terms, we seem to have a paradigmatic example of the mechanist Darwin clashing with the organicist Wallace. It is true that Wallace was an enthusiast for natural selection, but, as we have seen, he always thought of this in holistic terms. And the appeal to spirit forces reminds one of Aristotelian forces, which in Chapter 8 we shall see returning in France in a major way influencing organismic theologies. For now, while sensitive to dissenters, we will continue with the supposition that Darwinian theory is under the machine metaphor.

In respects, Darwin's book is unremarkable and, as we shall see when we come to things such as sexual dimorphism, much what one might have expected from the rich, upper-middle-class, white, male Englishman who had authored the *Origin*. We start with some general reflections on the plausibility of our evolutionary origins, followed by discussion of our differences from the apes. We must "admit that there is a much wider interval in mental power between one of the lowest fishes, as a lamprey or lancelet, and one of the higher apes, than between an ape and man" (Darwin 1871, 1, 35). Expectedly, natural selection has a major role to play here and, whenever convenient, Wallace's thinking was compared unfavorably. As is flagged by the full title of the work, the nature and effects of Darwin's secondary mechanism, sexual selection, got a detailed and comprehensive treatment. This was spurred by Wallace's claim that such things as human hairlessness can hardly be the consequence of natural selection. Darwin agreed with the point Wallace was making; but, rather than turning to spirit forces for help, he argued that sexual selection rather than natural selection was the cause: "The absence of hair on the body is to a certain extent a secondary sexual character; for in all parts of the world women are less hairy than men. Therefore we may reasonably suspect that this is a character which has been gained through sexual selection" (Darwin 1871, 2, 376). Humans are animals and as such subject to the same laws as other animals. Light is indeed thrown on man and his history.

Philosophy (Epistemology)

What then of our theory of knowledge? Does it make a difference to learn that we are not the climax of God's creative efforts on the Sixth Day? Are our knowledge claims and their foundations unchanged when we learn we are the end-product of a slow, natural, unguided process of selection? Start with Darwin himself. He was a scientist not a philosopher, but he had had a good classical education, so at once he thought in those terms when trying to see just how evolution would affect our thinking. "Plato Erasmus says in Phaedo that our 'necessary ideas' arise from the preexistence of the soul, are not derivable from experience.—read monkeys for preexistence—" (Darwin 1987, M 128; the Erasmus referred to here is Darwin's older brother). This is the key to human knowledge. Our brains, to use the modern analogy, are computers but they are programmed by natural selection to think in certain ways rather than other ways. Buy into Immanuel Kant's theory of knowledge. We think in certain ways rather than other ways because of the a priori categories of the understanding. These are in the mind before we start. However, for a Darwinian, they are not really a priori. They are Darwinian adaptations, because those of our would-be ancestors who had them – had them not by design but by the regular processes of selection on random variations – survived and reproduced, and those of our would-be ancestors who did not have them did not survive and reproduce. What's the point of thinking as we do rather than some other way? What's the final cause of our thought processes? Success in the struggle for existence.

There's no magic. Mathematics. Those hominins – proto-humans – who saw two bears go into a cave and only one come out and who then decided to sleep somewhere else have successors today. Those hominins – proto-humans – who saw two bears go into a cave and only one come out, and who then said "let's get out of the rain," are conspicuously absent when asked about what happened when they went into the cave (Ruse 1986, 2021a). Science. We have encountered the consilience of inductions. It is vital in science, but it is not confined to science. Sherlock Holmes looks for the bloodstains and the method of killing and the opportunity and the motive and the broken alibi and declares: "Professor, I accuse you of murdering your junior colleague, because she was about to reveal that you have not been contributing your share to the coffee fund." Same back then. At the end of a long day hunting and gathering, you finally get to the pool where you can have a nice, long, cool drink. You see footprints in the mud. The bushes are trampled on. There is a sound of growling in the undergrowth. You say, "I'm not really that thirsty," and move on quickly. You will live to hunt and gather another day. You say, "Tigers? Just a theory not a fact. Nothing to worry about." Don't bank on tenure.

Pragmatism

Did any professional philosophers buy into this line of thought? We must cross over to the New World. The American Charles Sanders Peirce openly embraced Darwin's approach:

> Not man merely, but all animals derive by inheritance (presumably by natural selection) two classes of ideas which adapt them to their environment. In the first place, they all have from birth some notions, however crude and concrete, of force, matter, space, and time; and, in the next place, they have some notion of what sort of objects their fellow-beings are, and of how they will act on given occasions. (Peirce 1883, 178)

Adding: "Side by side, then, with the well established proposition that all knowledge is based on experience, and that science is only advanced by the experimental verifications of theories, we have to place this other equally important truth, that all human knowledge, up to the highest flights of science, is but the development of our inborn animal instincts" (180–81).

Pragmatism! In a series of articles published in *Popular Science Monthly* between November 1877 and August 1878, Peirce spelt things out in terms of a "habit." "That which determines us, from given premises, to draw one inference rather than another, is some habit of mind, whether it be constitutional or acquired" (Peirce 1877, 3). Habits are up-to-date versions of Kant's categories of the understanding (Murphey 1968). Take causation and think of how we determine causation through consilience. For Kant, in some way, our mind is predetermined to guide us into certain sorts of thinking rather than others. Beware tigers, rather than tigers just a hypothesis. Naturalized, these come as habits. You think or reason this way rather than another way. But what determines the choice? Not some kind of disinterested, necessary conditions of thinking. It is, rather, the practical consequences. "The essence of belief is the establishment of a habit; and different beliefs are distinguished by the different modes of action to which they give rise." Spelling this out: "Consider what effects, that might conceivably have practical bearings, we conceive the object of our conception to have. Then, our conception of these effects is the whole of our conception of the object" (Peirce 1958, 5.402). Understand, it is not the consequences that are true or false. They happen or don't happen. It is the belief, informed by habit, that is true or false. It is the consequences that determine what is true or false. And what is crucially important is that this is a group effort: "reality is independent, not necessarily of thought in general, but only of what you or I or any finite number of men may think about it" (Peirce 1878, iv). Group effort, but not group selection. You can benefit from what others think or thought. But note that it is you who is benefiting, and the group only as a function of individuals.

Peirce was not alone in thinking this way. A notable and worthy successor was William James (older brother of the novelist Henry James). In lectures given at the beginning of the twentieth century, explicitly acknowledging his debt to Peirce (who seems to have been as difficult a personality as James was warm and friendly), James defined his position, saying:

> [The pragmatic method] is to try to interpret each notion by tracing its respective practical consequences. What difference would it practically make to anyone if this notion rather than that notion were true? If no practical difference whatever can be traced, then the alternatives mean practically the same thing, and all dispute is idle. Whenever a dispute is serious, we ought to be able to show some practical difference that must follow from one side or the other's being right. (James 1907, 44)

And into the Next Century

Less enthusiasm was shown in Darwin's own country. Towards the end of his very long life, Bertrand Russell said openly: "My general outlook, in the early years of this century, was profoundly ascetic. I disliked the real world and sought refuge in a timeless world, without change or decay or the will-o'-the-wisp of progress" (Russell 1959, 155). As far as Pragmatism is concerned, he thought it little more than an excuse for grabbing and holding power. The weak will bend to the will of the strong. There are no outside checks.

> In all this I feel a grave danger, the danger of what might be called cosmic impiety. The concept of "truth" as something dependent upon facts largely outside human control has been one of the ways in which philosophy hitherto has inculcated the necessary element of humility. When this check upon pride is removed, a further step is taken on the road towards a certain kind of madness – the intoxication of power which invaded philosophy with Fichte, and to which modern men, whether philosophers or not, are prone. (Russell 1945, 827)

Pragmatism fails both epistemologically and ethically.

Bertrand Russell set the tone for the twentieth century. Analytic philosophy, the dominant approach, wanted nothing to do at all with evolutionary thinking. Hugely influential was the Austrian-born philosopher Ludwig Wittgenstein. In his *Tractatus* he said: "The Darwinian theory has no more to do with philosophy than has any other hypothesis of natural science" (Wittgenstein 1922, 4.1122). No more to do with philosophy than Boyle's law! The English translation appeared in 1922, with a fulsome introduction by Russell: "to have constructed a theory of logic which is not at any point obviously wrong is to have achieved a work of extraordinary difficulty and importance. This merit, in my opinion, belongs to Mr Wittgenstein's book, and makes it one which no serious philosopher can afford to neglect." Twenty years later, Wittgenstein was still showing an indifference bordering on contempt. Now Darwinism is not merely judged irrelevant, but incorrect into the bargain. "I have always thought that Darwin was wrong: his theory doesn't account for all this variety of species. It hasn't the necessary multiplicity" (Rhees 1981, 174). As always, gnomic remarks like this come with absolutely no hint that the speaker has ever glanced at the writings of professional evolutionists.

In the first half or more of the twentieth century, there was significant non-Darwinian overlap between the philosophies of Britain and America. Thanks to his American disciple Norman Malcolm, Cornell University was more Wittgenstein-enthusiastic than Cambridge. Yet the Peirce–James approach still throve in certain American quarters, if not necessarily philosophy departments. The third of the great Pragmatists – John Dewey – enters the story. His philosophy was more interactive than that of the earlier two. Knowledge is not just a matter of what we do and think, but of the dialogue between what is in there (us) and what is out there (the world). "Truth, in final analysis, is the statement of things 'as they are,' not as they are in the inane and desolate void of isolation from human concern, but as they are in a shared and progressive experience..." (Dewey 1910, in Dewey 1976: 67; 118).

Where Dewey had his greatest influence was in his application of his philosophy to education.

> The educational point of view enables one to envisage the philosophic problems where they arise and thrive, where they are at home, and where acceptance or rejection makes a difference in practice. If we are willing to conceive education as the process of forming fundamental dispositions, intellectual and emotional, toward nature and fellow-men, *philosophy may even be defined as the general theory of education.* (Dewey 1916, in Dewey 1976: 338; my italics)

We start with the basic truth: "The native and unspoiled attitude of childhood, marked by ardent curiosity, fertile imagination, and love of experimental inquiry, is near, very near, to the attitude of the scientific mind" (Dewey 1910, in Dewey 1976: 179). Then, we use or encourage this eagerness to entangle with the outside world in the way we teach and the way children learn. Teaching science, for instance, is a matter of hands-on experimenting rather than copying truths being written out on the blackboard. As Herschel said, whirling a piece of stone at the end of a piece of string. Getting the feeling of the force right there in your hand. There is, no doubt, the danger that unstructured experience leads to sterile relativity; to the 1960s mantra: "If it feels good, then it is good." To avoid this false trail, the teacher must bring to the classroom much that humans have discovered already, and this must help structure the direction of inquiry and learning. Education is "continually shaping the individual's powers, saturating his consciousness, forming his habits, training his ideas, and arousing his feelings and emotions. Through this unconscious education the individual gradually comes to share in the intellectual resources which humanity has succeeded in getting together." In short: "He becomes an inheritor of the funded capital of civilization. The most formal and technical education in the world cannot safely depart from this general process. It can only organize it; or differentiate it in some particular direction" (Dewey 1897, 77). The truly great teacher, with ability and experience, can balance the two – hands-on experience versus direction based on what we already know. All of us who have long been teachers have met the rare Bobby Charlton, the great soccer player, whose brilliance in balancing makes one feel very humble.

Back in the realm of pure philosophy, in America, while it can hardly be said that it became a dominant position, Pragmatism continued to have its supporters. The major philosopher W. V. O. Quine was asked about the problem of induction – how and why one can infer regularities from single cases:

> Why there have been regularities is an obscure question, for it is hard to see what would count as an answer. What does make sense is the other part of the problem of induction: why does our innate subjective spacing of qualities accord so well with the functionally relevant groupings in nature as to make our inductions tend to come out right? Why should our objective spacing of qualities have a special purchase on nature and a lien on the future? (Quine 1969, 126–27)

He answered his own question: "There is some encouragement in Darwin. If people's innate spacing of qualities is a gene-linked trait, then the spacing that has made for the most successful inductions will have tended to predominate through natural selection. Creatures inveterately wrong in their inductions have a pathetic but praiseworthy tendency to die before reproducing their kind" (Quine 1969, 48).

A few others wrote in a similar manner. The best-known of them all, Richard Rorty, was adamant that we humans are at one with the rest of the living world, and that this is a consequence of being a Darwinian: "Darwinism requires that we think of what we do and are as continuous with what amoebas, spiders, and squirrels do and are" (1998, 295). And he was openly a Pragmatist about all of this, with John Dewey as his hero. He wrote that Pragmatists "should see themselves as working at the interface between the common sense of their community, a common sense much influenced by Greek metaphysics and monotheism, and the startlingly counterintuitive self-image sketched by Darwin, and partially filled in by Dewey" (41).

This was not a universal opinion. Back in England, the important philosopher of science, Karl Popper, like Wittgenstein Austrian-born, did not think that Darwin's theory was genuine science! He thought rather that it was a "Metaphysical Research Programme":

> Take "Adaptation". At first sight natural selection appears to explain it, and in a way it does, but it is hardly a scientific way. To say that a species now living is adapted to its environment is, in fact, almost tautological. Indeed we use the terms "adaptation" and "selection" in such a way that we can say that, if the species were not adapted, it would have been eliminated by natural selection. Similarly, if a species has been eliminated it must have been ill adapted to the conditions. Adaptation or fitness is defined by modern evolutionists as survival value, and can be measured by actual success in survival: there is hardly any possibility of testing a theory as feeble as this. (Popper 1974, 137)

We have here a classic instance of a philosopher holding forth on a subject about which he knows little, and that which he does know is wrong. To correct. Natural selection is not the tautology: "that which survives is that which survives." It is rather: "That which survives in one instance will, all other things being equal, probably survive in another instance." The only thing metaphysical is Popper's analysis of the topic (Ruse 1977).

To be fair, Popper took note of criticisms and grew to acknowledge the virtues of Darwin's theory, thereby confirming his famous dictum: "We learn by our mistakes." At least Popper was always one up on Alvin Plantinga, the distinguished Calvinist philosopher of religion. He does not much care for evolution at all! Plantinga agrees that, if we are Darwinian evolutionists, then this must extend to our reasoning and cognitive powers. But, he happily points out, Darwinian evolution cares nothing for truth, only for survival and reproductive success. Hence, there is no expectation of our reasoning and cognitive powers telling us the truth about the world: They just tell us what we need to believe to survive and reproduce, which information (although effective) could as easily be quite false. Plantinga tells the story of an overly rich dinner in an Oxford College, where Richard Dawkins spoke up for atheism before the philosopher A. J. Ayer – a classic case of preaching to the converted, I should have thought – and then goes on to draw a philosophical moral. Perhaps none of our thoughts can tell us about reality. Perhaps we are like beings in a dream world:

> Their beliefs might be like a sort of decoration that isn't involved in the causal chain leading to action. Their waking beliefs might be no more causally efficacious, with respect to their behaviour, than our dream beliefs are with respect to ours. This could go by way of pleiotropy: genes that code for traits important to survival also code for consciousness and belief; but the

> latter don't figure into the etiology of action. It *could* be that one of these creatures believes that he is at that elegant, bibulous Oxford dinner, when in fact he is slogging his way through some primeval swamp, desperately fighting off hungry crocodiles. (Plantinga 1993, 223–24)

Everything we believe about evolution could be false, and if that is not a refutation of Darwinian epistemology, nothing is! Plantinga (rather cutely) refers to this as "Darwin's Doubt" because the aged Darwin himself expressed worries of this kind: "With me the horrid doubt always arises whether the convictions of man's mind, which have been developed from the mind of the lower animals, are of any value or are at all trustworthy. Would anyone trust in the convictions of a monkey's mind, if there are any convictions in such a mind?" (Plantinga 1993, 219, quoting Darwin 1887, 1, 315–16). Plantinga does not mention that Darwin immediately excused himself as a reliable authority on such philosophical questions, but no matter.

Candidly, I am not sure how seriously we are supposed to take Plantinga's argument and example. It is certainly true that even the most ardent Darwinian agrees that there are times when organisms and their characteristics will be out of adaptive focus. Genetic drift would be a case in point, as would be that phenomenon mentioned by Plantinga, "pleiotropy," where a single gene controls two different characteristics and a nonadaptive feature piggybacks on an adaptive feature. But in no sense does Plantinga describe a situation remotely like the way that evolution truly works. Something like drift has minor effects – effects so minor (by definition) that they can slip under selection. Thinking that you are boozing it up with Freddie Ayer is not the way to fight off crocodiles. Fighting crocodiles requires force and defence and cunning and fear and split-second reaction to danger and much more. Go down to the Florida Everglades if you doubt this. At the very most, things such as drift and pleiotropy and so forth are going to make one a bit uncertain about some things occasionally: when those things do not much matter anyway. If we need to know the truth, and we do need to know the truth when faced with crocodiles, drift will not stand in our way. It is not sufficiently powerful to make evolution through selection that maladaptive or ineffective.

Plantinga is clearly motivated by his religious beliefs. Let us therefore leave further discussion until we get to Chapter 8 on religion. We have one last topic on epistemology to discuss.

Cultural Evolution

Many, probably a majority, of those who call themselves "evolutionary epistemologists," have been pursuing a track not yet discussed, namely focusing on *cultural* evolution rather than *biological* evolution. Thus far we have been thinking in terms of biological evolution – how the natural selection of some animals over others has led to different ways of thinking, some more effective – better adaptations – than others. Cultural evolution, rather, remains at the level of concepts and their relationships, looking at the way they change and the reasons for such change. T. H. Huxley was an early enthusiast: "The struggle for existence holds as much in the intellectual as in the physical world. A theory is a species of thinking, and its right to exist is coextensive with its power of resisting extinction by its rivals" (Huxley 1880, 15–16). This is a claim made implicitly and supported by Darwin himself. In the *Descent of Man*, Darwin wrote of the evolution of language in these terms. "Languages, like organic beings, can be classed in groups under groups; and they can be classed either naturally according to descent, or artificially by other characters. Dominant languages and dialects spread widely and lead to the gradual extinction of other tongues" (1871, 1, 60). There are other similarities with the biological: "The same language never has two birth-places." But then differences start to appear:

> Distinct languages may be crossed or blended together. We see variability in every tongue, and new words are continually cropping up; but as there is a limit to the powers of the memory, single words, like whole languages, gradually become extinct. As Max Müller has well remarked: – "A struggle for life is constantly going on amongst the words and grammatical forms in each language. The better, the shorter, the easier forms are constantly gaining the upper hand, and they owe their success to their own inherent virtue." To these more important causes of the survival of certain words, mere novelty may, I think, be added; for there is in the mind of man a strong love for slight changes in all things. (1, 60)

Darwin ends by telling us: "The survival or preservation of certain favoured words in the struggle for existence is natural selection" (1, 60–61). This may be true, but talking in terms of the survival of words because of their "novelty" and how they charm their users is hardly the world of the *Origin of Species*.

Be this as it may, cultural evolution and its supposed implications for epistemology have proven to have staying power. One major enthusiast was the English historian and philosopher of science Stephen Toulmin (1967, 1972). He articulated a whole world picture of a culturally based science and its changes:

> Science develops as the outcome of a double process: at each stage, a pool of competing intellectual variants is in circulation, and in each generation a selection process is going on, by which certain of these variants are accepted and incorporated into the science concerned, to be passed on to the next generation of workers as integral elements of the tradition.

Continuing: "Moving from one historical context to the next, the actual ideas transmitted display neither a complete breach at any point – the idea of absolute 'scientific revolutions' involves an over-simplification – nor perfect replication either." Hence:

> The change from one cross-section to the next is an *evolutionary* one in this sense too: the later intellectual cross-sections of a tradition reproduce the content of their immediate predecessors, as modified by those particular intellectual novelties which were selected out in the meanwhile – in the light of the professional standards of the science of the time (1967, 465–66).

A nice picture perhaps, but the critic will wonder how Darwinian? For a start, the new variations (mutations) in the Darwinian world are not directed. As Darwin's eighteen-month (cultural) search for the cause of evolution shows clearly, it was very much directed. With natural selection as its endpoint! Darwin knew what he wanted. Darwin got what he wanted. For a second, as with Darwin's own language example, the way in which two lines blend is not the way that species do in nature. An alligator and a cow are never going to produce offspring, and, in those cases of breeding between close species, as often as not the offspring are problematic: the mule, for example. To the contrary in science, the coming together of different fields, as in the Whewellian consilience of the *Origin*, is the sign of a major breakthrough. Not just a major breakthrough, but a major *progressive* breakthrough. As Karl Popper argues, the process of science may never end, but it does improve. Copernicus over Aristotle, Huygens over Newton, Darwin over Sedgwick and company.

Other pieces of evidence have supposedly supported cultural evolution, for example "horizontal gene DNA transfer" from one species to a different one. Whether putting a little piece of DNA in another species would be equivalent to combining paleontology, embryology, and biogeography into one theory is hardly yet proven. There have been suggestions that of the 20,000 total genes that produce humans, up to 100 came through gene transfer, although these kinds of figures have been strongly challenged (Salzberg et al. 2001). What one can say is that, if such transfer generally occurred in a significant fashion, one would expect to find that the tree of life has been replaced by a bush of life. This seems not to be the overall case (Theobald 2010). There is no evidence that a horse and a bird got together and the result was Pegasus. Likewise, no evidence that a human and a bird got together and the result was the Archangel Michael.

It does nevertheless seem, despite the obvious difficulties, that the idea of cultural evolution has proven to be rather fitter than one might have expected. A variant was started by Richard Dawkins (1976) in his *Selfish Gene*. There he supposes that equivalent to the gene in the biological world there is the "meme" in the cultural world. Some philosophers have taken up this idea with enthusiasm, most notably the well-known philosopher Daniel Dennett (1995). Few apparently shared his enthusiasm. *The Journal of Memetics* lasted less than ten years. But let us not end this discussion on an entirely negative note. The cultural approach may not yield all the benefits its supporters claim for it, but it can be highly instructive in a heuristic fashion, leading to new insights and understanding. It is just that it cannot be taken too literally – especially in a world where Thomas Kuhn, he of incommensurable paradigms, tells us that his overall approach is evolutionary!

Philosophy (Ethics)

In discussing morality, philosophers traditionally make a distinction between two levels of inquiry. First, there is "substantive" or "normative" ethics: What should I do? Second, there is "metaethics": Why should I do that which (substantive-ethics-guided) I do? Christianity makes the distinction clear. What should I do? "Love my neighbor as myself." Why should I do that which I do? "Because that is what God wants." The same questions come up when we are discussing Darwinism. What does Darwinism tell me I should do? Why does Darwinism tell me what I should do?

Traditionally, it is believed that, at the substantive level, the theory of evolution through natural selection directs us to what is known as "Social Darwinism" (O'Connell and Ruse 2021). Nature is red in tooth and claw. That is what we humans should emulate. Spencer is the classic case. As far as behavior is concerned, he is blunt:

> We must call those spurious philanthropists, who, to prevent present misery, would entail greater misery upon future generations. All defenders of a Poor Law must, however, be classed among such. That rigorous necessity which, when allowed to act on them, becomes so sharp a spur to the lazy and so strong a bridle to the random, these paupers' friends would repeal, because of the wailing it here and there produces. (Spencer 1851, 323)

This was the clarion call that soon attracted supporters from many lands and professions. Typical were the views of Andrew Carnegie, Scottish-born owner of massive steelworks in Pittsburgh, Pennsylvania, and the force behind the crushing of the workers in the Homestead Strike of 1892: "The price which society pays for the law of competition, like the price it pays for cheap comforts and luxuries, is also great; but the advantages of this law are also greater still, for it is to this law that we owe our wonderful material development, which brings improved conditions in its train." Adding that "while the law may be sometimes hard for the individual, it is best for the race, because it insures the survival of the fittest in every department" (Carnegie 1889, 655).

What metaethical justification was offered to justify these harsh sentiments? Spencer was explicit: progress. Since nature is an organic unfolding, getting ever-more perfect, prescriptions emerge naturally. We ought to cherish the evolutionary process as generating ever-greater value, and hence we ought to help it along. At least, not impede its progress. "Ethics has for its subject-matter, that form which universal conduct assumes during the last stages of its evolution" (Spencer 1879, 21). Adding: "And there has followed the corollary that conduct gains ethical sanction in proportion as the activities, becoming less and less militant and more and more industrial, are such as do not necessitate mutual injury or hindrance, but consist with, and are furthered by, co-operation and mutual aid" (Spencer 1879, 21). As one unpacks what this means and entails, one starts to wonder if one has told the whole story about the harsh prescriptions of Social Darwinism. Reflecting the Quaker influences of his childhood, Spencer was strongly against militarism and bemoaned the naval arms race which, by the end of the century, absorbed Britain and Germany. Likewise, Carnegie is famous for saying that no man should die rich. Putting this belief into practice, as is well known, he used his fortune to spark the founding of public libraries. This author was one who benefited, spending many happy and fruitful hours in the Carnegie Library of Walsall, a town in the British Midlands (with lots of rain!). And indeed, as one digs further, one finds that it is too easy to misinterpret what is truly being said by the Social Darwinians. Spencer had nothing against widows and children. His gripe was against those in power who refused to let the gifted yet poor rise up in their societies. As with Margaret Thatcher a century later – like Spencer from a lower-middle-class, nonconformist background – he wanted to abolish practices that led people not to make the needed effort but rather stay at the bottom thanks to state support.

Darwin's Thinking on Morality

Prima facie, T. H. Huxley seemed to be a tailor-made Social Darwinian. In a well-known, late-life essay on evolution and ethics, entirely ignoring his earlier hesitations about the causal efficacy of natural selection, he argued: "Man, the animal, in fact, has worked his way to the headship of the sentient world, and has become the superb animal which he is, in virtue of his success in the struggle for existence." He "has been largely indebted to those qualities which he shares with the ape and the tiger; his exceptional physical organization; his cunning, his sociability, his curiosity, and his imitativeness; his ruthless and ferocious destructiveness when his anger is roused by opposition." However, this is tempered and controlled by our moral sense. Here Huxley relied on the analogy between cultural evolution and biological evolution. Morality is entirely a cultural phenomenon: "Of moral purpose I see not a trace in nature. That is an article of exclusively human manufacture" (Huxley 1900, 2, 285). Our biological self is curbed by our cultural self: "Ethical nature may count upon having to reckon with a tenacious and powerful enemy as long as the world lasts. But, on the other hand, I see no limit to the extent to which intelligence and will, guided by sound principles of investigation, and organized in common effort, may modify the conditions of existence, for a period longer than that now covered by history" (Huxley 1893a, 36).

Where did Darwin stand on all of this? One can find Social Darwinian sentiments in Darwin's writings, especially if one's gaze lifts from the publications to the less formal and more candid letters that he wrote. Responding to a book, *The Creed of Science: Religious, Moral, and Social*, Darwin wrote (to the author): "I could show fight on natural selection having done and doing more for the progress of civilisation than you seem inclined to admit." How come? "Remember what risk the nations of Europe ran, not so many centuries ago of being overwhelmed by the Turks, and how ridiculous such an idea now is in more civilised so-called Caucasian races have beaten the Turkish hollow in the struggle for existence." In the light of this: "Looking to the world at no very distant date, what an endless number of the lower races will have been eliminated by the higher civilised races throughout the world" (Letter to William Graham, July 3, 1881).

As always when Darwin is writing on these matters – confirming what we have seen so often about the iron grip of social status in the Victorian era – these truly are the words of someone we have met already: an upper-middle-class Englishman, writing at the peak of the British Empire, rather than a serious scientist. When we look at what he wrote about morality in the *Descent*, where he was writing in a descriptive rather than prescriptive mode, his position is much more subtle. As with epistemology, Darwin thought that more conventional morality – love your neighbor as yourself, that sort of thing – came about because of rather than despite (biological) natural selection (Ruse 2022a, b). Tribes of people who get along and help each other do better than tribes who don't:

> It must not be forgotten that although a high standard of morality gives but a slight or no advantage to each individual man and his children over the other men of the same tribe, yet that an advancement in the standard of morality and an increase in the number of well-endowed men will certainly

> give an immense advantage to one tribe over another. There can be no doubt that a tribe including many members who, from possessing in a high degree the spirit of patriotism, fidelity, obedience, courage, and sympathy, were always ready to give aid to each other and to sacrifice themselves for the common good, would be victorious over most other tribes; and this would be natural selection. (Darwin 1871, 1, 166)

"Victorious over most other tribes"? Is this not an appeal to group selection? Not at all! Shortly before this passage, Darwin implies that (what today is known as) "reciprocal altruism" is a major causal factor. You scratch my back and I will scratch yours: "as the reasoning powers and foresight of the members [of a tribe] became improved, each man would soon learn from experience that if he aided his fellow-men, he would commonly receive aid in return" (1, 163). Furthermore, Darwin thought that what we now call "kin selection" was at work. As with the sterile Hymenoptera, members of tribes are interrelated or think they are – the founding ancestor being the Wolf or some such body.

That we have this moral sense explains some of the more convoluted arguments in the *Descent*. There are passages that make Herbert Spencer look like a wimp:

> With savages, the weak in body or mind are soon eliminated; and those that survive commonly exhibit a vigorous state of health. We civilised men, on the other hand, do our utmost to check the process of elimination; we build asylums for the imbecile, the maimed, and the sick; we institute poor-laws; and our medical men exert their utmost skill to save the life of every one to the last moment. There is reason to believe that vaccination has preserved thousands, who from a weak constitution would formerly have succumbed to small-pox. (1, 168)

With obvious bad consequences.

> Thus the weak members of civilised societies propagate their kind. No one who has attended to the breeding of domestic animals will doubt that this must be highly injurious to the race of man. It is surprising how soon a want of care, or care wrongly directed, leads to the degeneration of a domestic race; but excepting in the case of man himself, hardly any one is so ignorant as to allow his worst animals to breed. (1, 168)

But things are not quite this simple.

> The aid which we feel impelled to give to the helpless is mainly an incidental result of the instinct of sympathy, which was originally acquired as part of the social instincts, but subsequently rendered, in the manner previously

> indicated, more tender and more widely diffused. Nor could we check our sympathy, if so urged by hard reason, without deterioration in the noblest part of our nature. (1, 168–69)

It seems to be a case of "damned if you do, damned if you don't." We can say that, given his provenance, it would be odd indeed if Darwin did not share some of the social prejudices of his class. The Wedgwoods made their fortune by treating their workers, all 15,000 of them, like machines. But also, given his provenance, it would be odd indeed if Darwin, as squire of his village in all but name, felt no compassion or duty to those in need. This is the man who contributed to the Downe coal and clothing club for the needy in winter. (For several years, he was the treasurer.) Darwin felt the tug each way, and so we can readily understand why he found that natural selection pointed to reasons that we have this conflict. Prima facie this all sounds a bit like Huxley twenty years later. But Huxley thinks our moral sense is entirely cultural, whereas Darwin thinks it is a biological, selection-caused adaptation. Very different positions. For Huxley, morality is rational. For Darwin, there is absolutely no overriding reason why substantive ethics should be consistent. What works is what works.

What about metaethical justification? Darwin was a scientist not a philosopher, but his position was clear. There is no justification for morality beyond its being of adaptive significance for human beings. In other words, his position was essentially that of Pragmatism transferred to the realm of morality. Darwin is what is known as a "nonrealist," denying that there are external, objective, moral laws. If they worked, we could well believe and behave on very different directives:

> I do not wish to maintain that any strictly social animal, if its intellectual faculties were to become as active and as highly developed as in man, would acquire exactly the same moral sense as ours. In the same manner as various animals have some sense of beauty, though they admire widely different objects, so they might have a sense of right and wrong, though led by it to follow widely different lines of conduct. If, for instance, to take an extreme case, men were reared under precisely the same conditions as hive-bees, there can hardly be a doubt that our unmarried females would, like the worker-bees, think it a sacred duty to kill their brothers, and mothers would strive to kill their fertile daughters; and no one would think of interfering. Nevertheless the bee, or any other social animal, would in our supposed case gain, as it appears to me, some feeling of right and wrong, or a conscience. (Darwin 1871, 1, 73)

We are humans not hive-bees and, thank God or Darwin, we males need not worry as winter approaches. But the relativity point is made. It is hard to talk of "progress" when we might have functioned well as honeybees. Objective standards are shown to be impossible. To use a term popular among philosophers, they have been "debunked" (Wielenberg 2010).

Reactions

Judging by Bertrand Russell's reaction to Pragmatism, looking at British professional philosophers, we should not anticipate huge enthusiasm about a Darwin-influenced approach to morality. Good advice, because the sentiments were much akin to the feelings about epistemology, and for similar underlying reasons. As with Russell, so likewise his counterpart in moral philosophy, G. E. Moore – illustrating how out of touch were the echelons of the top levels of English higher education from the ever-active, yet dirty and slum-filled world that the country had become – agreed that the rules of morality exist in a kind of ethereal world, as do the rules of mathematics. "I am pleased to believe that this is the most Platonic system of modern times" (Baldwin 1990, 50).

With the death of Mill in 1873, the leading British moral philosopher was Henry Sidgwick, who held forth on evolution and ethics in 1876 in the newly founded journal *Mind*. He accepted the fact of evolution and even that our moral faculty was part of the picture. From then on, it was downhill. The truth must now be told: "It is more necessary to argue that the theory of Evolution, thus widely understood, has little or no bearing upon ethics" (Sidgwick 1876, 54). Expectedly, it is Spencer – the man of letters – rather than Darwin – the man of science – who is the focus of Sidgwick's attack. Spencer's claim that moral excellence demands ever-more-advanced evolutionary states is found wanting. "'Slowly but surely,' writes Mr. Spencer, 'Evolution brings about an increasing amount of happiness,' so that we are warranted in believing that 'Evolution can only end in the establishment of the most complete happiness.'" Nonsense, responds Sidgwick: If "we confine ourselves to human beings, to whom alone the practical side of the doctrine applies, is it not too paradoxical to assert that 'rising in the scale of existence' means no more than 'developing further the capacity to exist'?" It all adds up: "A greater degree of fertility would thus become an excellence outweighing the finest moral and intellectual endowments; and some semi-barbarous races must be held to have attained the end of human existence more than some of the pioneers and patterns of civilization" (59).

We come to Spencer's student, Moore, the author of the influential *Principia Ethica*. (When I was an undergraduate in the early 1960s, *Principia Ethica* was still the text from which we were taught.) If Sidgwick could be said to have focused on the substantive ethical side of things – evolution does not produce people doing good as we understand it – Moore's focus was on metaethics. He found his cue in David Hume, who famously said you cannot get "ought" statements from "is" statements:

> In every system of morality, which I have hitherto met with, I have always remarked, that the author proceeds for some time in the ordinary way of reasoning, and establishes the being of a God, or makes observations concerning human affairs; when of a sudden I am surprized to find, that instead of the usual copulations of propositions, is, and is not, I meet with no proposition that is not connected with an ought, or an ought not. This change is imperceptible; but is, however, of the last consequence. For as this ought, or ought not, expresses some new relation or affirmation, it is necessary that it should be observed and explained; and at the same time that a reason should be given, for what seems altogether inconceivable, how this new relation can be a deduction from others, which are entirely different from it. (Hume 1739–40, 469)

Moore's version of this he called the "naturalistic fallacy." If a man "confuses 'good,' which is not in the same sense a natural object, with any natural object whatever, then there is a reason for calling that a naturalistic fallacy; its being made with regard to 'good' marks it as something quite specific, and this specific mistake deserves a name because it is so common" (Moore 1903, 14). Herbert Spencer is a blatant sinner: "He tells us that one of the things it has proved is that conduct gains ethical sanction in proportion as it displays certain characteristics. What he has tried to prove is only that, in proportion as it displays those characteristics, it is more evolved" (Moore 1903, 31). Alas, "more evolved" is a matter of fact. "Conduct gains ethical sanction" is a matter of obligation. Hume put his finger on the problem here. He critiqued works that assume there is a deductive connection.

Evolutionary ethics fails both at the substantive level and at the metaethical level.

The Twentieth Century

This is not to deny that through the early decades of the twentieth century, although very much a minority interest, often promoted by biologists rather than philosophers, one did see ongoing attempts to present a plausible evolutionary ethics. The Spencer/Darwin divide continued. One who fell firmly on the Spencerian side was, expectedly, given his enthusiasm for progress, the already introduced Julian Huxley (1943), grandson of T. H. Huxley. As expected, C. D. Broad, student of Moore, jumped all over Huxley and his arguments:

> Whilst I agree that a knowledge of the facts and laws of evolution might have considerable and increasing relevance to the question whether certain acts would be right or wrong, since it might help us to foresee the large-scale and long-range consequences of such acts, I am unable to see that it has any direct bearing on the question whether certain states of affairs or processes or experiences would be intrinsically good or bad. (Broad 1944, 367)

The naturalistic fallacy throve as much during the Second World War as it did during the Boer War.

Not that this dissuaded later generations of Spencer-influenced biologists. In 1975, at the beginning of *Sociobiology: The New Synthesis*, Edward O. Wilson rode roughshod over the philosophers: "Scientists and humanists should consider together the possibility that the time has come for ethics to be removed temporarily from the hands of the philosophers and biologicized" (Wilson 1975, 3). You can imagine how this went down in the analytic philosophical community. Wilson and I cowrote a paper on the topic (Ruse and Wilson 1986). It was much anthologized as an example of how not to do philosophy! Not that criticisms were about to stop either of us. Wilson's prescriptions reflected the challenges of our era. He had concern about the environment, specifically about biodiversity (Wilson 1984, 1992, 2012). This was expressed through his "biophilia" hypothesis. "To explore and affiliate with life is a deep and complicated process in mental development. To an extent still undervalued in philosophy and religion, our existence depends on this propensity, our spirit is woven from it, hope rises on its currents" (Wilson 1984, 1). In the organicist tradition, Wilson saw all of life as a value-laden, interconnected whole. Individual organisms, individual species, are part of a larger network and no one or group can take itself apart in isolation. Morally, therefore, our obligation is to preserve life. Not life, as such, in its own right. Life as pointing to and supporting the well-being of humans. One preserves plants in the Amazon because they may turn out to have medicinal qualities of value to our species.

The leading Anglophone moral philosopher of the second half of the twentieth century was the American John Rawls. His basic (substantive ethical) claim was that justice is fairness. We ought to act so we do the best for everyone, and to achieve this we must put ourselves in the "original position" – not knowing where we find ourselves in society, privileged or otherwise, intelligent and attractive or otherwise – and then make our judgments, uninfluenced by personal considerations or profits. Where do we get these moral directives? We know that biological thinking, including Darwinian evolutionary theory, had always had a bigger input in American philosophy than in British. No great surprise then that Rawls appealed to Darwinism:

> In arguing for the greater stability of the principles of justice I have assumed that certain psychological laws are true, or approximately so. I shall not pursue the question of stability beyond this point. We may note however that one might ask how it is that human beings have acquired a nature described by these psychological principles. The theory of evolution would suggest that it is the outcome of natural selection; the capacity for a sense of justice and the moral feelings is an adaptation of mankind to its place in nature. (Rawls 1971, 502–3)

Rawls, like Plato in the *Republic*, was offering a contract theory – what's a good setup for a group of folk living together? One of the problems of such a theory is that, historically, it never seems very plausible that a group of wise men (and perhaps women) sat around and drew up the rules for proper conduct. It makes more sense to leave it to natural selection.

Still, an urge to be nice to small children is not a moral claim – you *ought* to be nice to small children. Rawls was fully aware of this. Even if biology does underlie contract theory, it doesn't follow that it is morally obligatory to follow it. "These remarks are not intended as justifying reasons for the contract view" (Rawls 1971, 504). Rather, we seem to have a Darwinized version of the so-called emotive theory of ethics (Ayer 1936). I don't like being unkind to small children. It upsets me. Naturally, therefore, I don't like it when you are unkind to small children. For my peace of mind, if not for yours, I urge you not to be unkind to small children. Does this mean that substantive morality is no more than an illusion put into place by our genes to make us good cooperators? An illusion which our biology leads us to objectify: You *ought* not be unkind to small children. Without this qualifying clause we would all cheat. Substantive morality is an adaptation with no more special standing than camouflage or sexual desire?

This is the Darwinian position and, without implying that this would have satisfied Rawls (who was more inclined to a Kantian a priori justification), there was an increasing number of professional philosophers drawn that way. One was the present author, who, when he cowrote with Wilson, was already moving to Darwinian nonrealism to the extent that he felt decisively queasy about putting his name to the paper. I had not lost my reason. Sidgwick had convinced the philosophical community that any attempt to link ethics with evolution is mistaken. In fact, the two positions are very different. No one (in the community) is denying the Hume defense of the is/ought barrier. Rather, the Darwinian does an end run around it. Evolution is not being used to *justify* morality, which is the position of the organicist, who sees values as built into the world. Rather, to *explain* why we have morality, which is the position of the mechanist, seeing the world as value-free. The modern-day Darwinian evolutionary ethicist accepts standard morality, as does Darwin; thinks there are psychological reasons why we have these beliefs, as does Darwin; and thinks we have these beliefs because it helps in the struggle for existence, as does Darwin (Ruse and Richards 2017). Humans who aid each other – who think they *should* aid others – even to the point of helping little old ladies (or little old men) across the road are going to be parts of groups who do better than groups who do not aid each other. (Groups succeeding but from the individual selection reason that we individually do better, or our relatives do better. The little old lady might be your mum and she helps babysit your kids. Or reciprocating with others who have mums and kids.) The key Darwinian component is that, since evolution is not progressive, we might have evolved in different ways and have different moralities. Remember the hive bees and the moral obligation to kick your brothers out of the nest as winter approaches. Moral obligation, but no objective foundations.

Point made.

Underlying Reasons

Why were the British so unresponsive to Darwinism? No big secret. In Victorian Britain, leading up to the turn of the century, idealism held sway. All influential was the Oxford philosopher T. H. Green, who explained everything in terms of a universal consciousness, something very similar to Plato's Form of the Good. "We must hold then that there is a consciousness for which the relations of fact, that form the object of our gradually attained knowledge, already and eternally exist; and that the growing knowledge of the individual is a progress towards this consciousness" (Green 1883, §69). As L. T. Hobhouse, Professor of Sociology at the University of London, wrote, the significance of evolutionary thinking stood not a chance in such an atmosphere. T. H. Green's influence "was dominant in Oxford and in the English and Scottish Universities generally in the Eighties and early Nineties. In this philosophy there seemed to many to be a way of escape not only from a barren individualism [implied by the struggle for existence] but from the whole philosophy of evolutionism" (Hobhouse 1913, xviii).

Wittgenstein and Popper should have done more homework before they dismissed Darwin; but remember that they started life in the world of German culture. General Friedrich von Bernhardi, pushed out of the German army because he was signaling a little too bluntly the General Staff's intentions, left no place for the imagination in his best-selling *Germany and the Next War* (1912, 18, 20). "War is a biological necessity," and hence: "Those forms survive which are able to procure themselves the most favourable conditions of life, and to assert themselves in the universal economy of nature. The weaker succumb" (18). Progress depends on war: "Without war, inferior or decaying races would easily choke the growth of healthy budding elements, and a universal decadence would follow" (20). For people who lived through the First World War – totally pointless but incredibly violent – one can understand why they rejected the theory that supposedly led to von Bernhardi's philosophy. This is to explain initial convictions. It is not to imply that they should never have changed their minds, as did Popper.

Crossing the Atlantic to America, we have Pragmatism, a philosophy that is much more evolution friendly, openly Darwinism friendly. One's initial inclination is to explain this difference between old and new as a function of the undoubted truth that America, after the Civil War, was a country springing ahead, with Carnegie's Social Darwinism ruling all. Not so. Peirce, for one, was very hostile to Social Darwinism, with its presumption that Darwinism generally applied to human societies:

> The gospel of Christ says that progress comes from every individual merging his individuality in sympathy with his neighbors. On the other side, the conviction of the nineteenth century is that progress takes place by virtue of every individual's striving for himself with all his might and trampling his neighbor under foot whenever he gets a chance to do so. This may accurately be called the Gospel of Greed. (Peirce 1893, 182)

Peirce felt so strongly on the subject that (not entirely consistently) by the 1890s he was running down natural-selection theory, arguing that it was not much of a theory, popular only because of its perceived support of the social philosophy of the day: "The extraordinarily favorable reception it met with was plainly owing, in large measure, to its ideas being those toward which the age was favorably disposed, especially, because of the encouragement it gave to the greed-philosophy" (Wiener 1949, 78).

To find the answer behind Peirce's Pragmatism, as always, we should look to past influences. All-important was New England transcendentalism, the neo-Romantic movement of people such as Ralph Waldo Emerson that owed so much to German thinkers, Kant and others, including Schelling. Somewhat ungraciously – although more likely playfully – Peirce wrote: "I am not conscious of having contracted any of that virus [transcendentalism]. Nevertheless, it is probable that some cultured bacilli, some benignant form of the disease was implanted in my soul, unawares, and that now, after long incubation, it comes to the surface, modified by mathematical conceptions and by training in physical investigations." Indeed. Just before this passage, Peirce restated the core thesis of his metaphysics: "I have begun by showing that *tychism* [the idea that chance plays a real role in the universe] must give birth to an evolutionary cosmology, in which all the regularities of nature and of mind are regarded as products of growth, and to a Schelling-fashioned idealism which holds matter to be mere specialized and partially deadened mind" (CP 6.102, Collected Papers of C. S. Peirce, quoted by Kruse 2010, 388).

In the words of a sensitive commentator:

> The idea that the human is part and parcel of nature is perhaps the single most characteristic idea in the American philosophies – transcendentalism, pragmatism, and naturalism – of the nineteenth and early twentieth centuries. In the thought of Charles Peirce, we find a synthesis of some of the core elements of Emersonian idealism into a rich theory of scientific inquiry that affirms the fundamental sympathy between mind and nature and commits itself to the vitality of the creative imagination in human inquiry. (Kruse 2010, 399–400)

When Darwin's theory arrived on the scene, it provided precisely the empirical understanding demanded by Peirce's mindset. That said, Peirce's heritage was decidedly Romantic – organicist – so there really is no great surprise that his relationship with Darwinism – mechanist – although crucial, was always somewhat fraught.

One further question that might be asked is why the British were never influenced by the Americans. The answer is depressingly obvious. The British have always looked down upon, condescended to, the Americans – especially after the Second World War, when it was so obvious that without American support and generosity the British were in line for a repeat of the Battle of Hastings. The food! The spelling! And a country that – horror of horrors – pronounces the last letter of the alphabet as "zee" rather than "zed"! Regrettably typical was the middle-of-the-century Oxford philosopher Gilbert Ryle, hugely influential – for many years he was editor of the leading philosophy journal *Mind* (where Sidgwick made his mark by concluding that evolution had nothing to do with understanding morality) – who declared that William James and John Dewey were the "Great American Bores" (Gibson 2023).

One might nevertheless ask if, independently, apart from responses to Americans, the mechanism/organism divide can be applied usefully to understand the British. After all, Moore claimed to be a Platonist and that surely applies to Russell also. As organicist as it comes. Whatever the roots of Peirce's views, Pragmatism fits more comfortably under mechanism. That is true, but whether focusing on the divide is helpful is another matter. Is the complete explanation of their differences to say the Americans tended towards mechanism and the British towards organicism? When dealing with science, understanding the physical world, the divide does play a role, although perhaps less than one might think. In their Spandrels paper, Gould and Lewontin were arguing openly (for whatever reason) from an organismic perspective, and the same is true of E. O. Wilson. One should not exaggerate. The overall tendency is towards mechanism, as is shown by the united opposition to Wilson's endorsement of group selection. Although there was the same in Britain, which does not sound very organismic.

Yet ask then a follow-up question: Was there no one in Britain interested in creating an empirical philosophy akin to, but not necessarily identical to, transcendentalism? A modern organicism that goes beyond Plato? There was! An ardent group of organicists, mainly continental but with English representation, notably J. H. Woodger in his *Biological Principles* (1929). Never one to beat about the bush, he was contemptuous of mechanism, speaking of it as "dogmatic" (249). Not so dogmatic, apparently, was the central organismic belief that the overall pattern of evolution bears striking analogy to the development of individual organisms, change starts within rather than without, and the overall pattern is progressive. Acorn to oak. Internal final cause, versus external. "A machine is made to realize some conscious human purpose. Its parts work together to secure that purpose, not to secure its own persistence. An organism is a mode of persistence" (436). Value assigned versus value discovered.

Note that taking an organismic position does not thereby preclude taking an evolutionary position. Indeed, as we know, with its emphasis on organic development it almost begs for an evolutionary interpretation. It precludes taking a Darwinian (and hence mechanistic) position. Expectedly, therefore, natural selection was dismissed quickly. Woodger pointed out that evolution depends on new variation: "any theory of evolution is driven to suppose change in the immanent factors." Unfortunately: "Natural selection supposes that they occur 'accidentally', i.e. it makes no assertions about the origin of the changes, and hence it is not strictly making a theory about the origin of species, but only about their survival" (403). Shades of Spandrels! It also turns out that the key factor about organisms is "organization," and the chief accomplishment of any adequate theory of evolution is explaining the increase in complexity of organization through time: "Thus the question of increase in organization is the crux of the whole problem." Natural selection fails again. "Natural selection will not help us at all because it only deals with the survival of immanent changes, not with their origin. It helps us to see how some attempts towards new modes of organization would not be successful after they had been reached, but not how they were reached" (413). There is more in this vein. Much more. Including a cutting dismissal of the work of Darwin himself:

> Charles Darwin has been called the Newton of biology, but it will be time enough to talk about the Newton of biology after our science has found its Galileo. To suppose that Darwin was the Newton of biology is to suppose that biology has already reached a degree of theoretical development comparable with that of physics in the eighteenth century, and that surely is preposterous. (483)

Few of the younger generation were convinced. As we move to what was characterized in Chapter 6 as the "Age of Normal Science," philosophers, led by David Hull (1969), started to see the merits of biology generally, including natural selection-driven evolution. Especially inasmuch as biology could be shown as getting more like the physical sciences, in the sense of obeying epistemic demands such as predictive fertility and falsifiability and consilience. Demands that, in *Biological Principles*, Woodger, on the grounds that they could not deal with complexity, explicitly eschewed. This is as may be. For the new cohort, there was little place, in the era of the double helix, for a philosophy that assured us confidently that it is deeply flawed "to study organisms piece by piece as the parts of a machine can be studied" (436)! How else did the molecular biologists break the genetic code other than by using the same techniques as were used, during the Second World War, by Alan Turing and others when they broke the code of the Enigma Machine at Bletchley Park? Woodger and his fellow organicists, thinking that they and only they had the key to understanding organisms, were living in cloud cuckoo land (Ruse 2021b). More particularly, they were staying at fancy rural retreats, having endless bull sessions about complexity and like topics (Nicholson and Gawne 2014). Meanwhile, Watson and Crick, when they could take their minds off birds and beer, were hunched over models built in their lab as they ferreted out the structure of the DNA model (Watson 1968).

Unsurprisingly, the new crop of philosophers of biology tended to ignore Woodger-type thinking. It simply wasn't looking at modern science. This contrast between organicist and mechanist approaches to evolution was certainly a motive force behind my *The Philosophy of Biology* (Ruse 1973), the theme of which was that the move of biology to be more physical science-like is supported by the incorporation of molecular findings (like the double helix) into biological discussions and the increasing resort to mathematics-infused thinking, as is shown by population genetics. With this movement, which my book sought to capture, it was not long before there were tentative steps to showing how biology (evolution particularly) offers foundations for both epistemology and ethics – or, more accurately, how biology shows that there cannot be traditional foundations. In another of the overviews that are, apparently, my forte – *Taking Darwin Seriously* (Ruse 1986) – I was one who argued for precisely this. This chapter has shown that such thinking was not always welcomed with enthusiasm and applause – as always with philosophers, there were those in the opposing, welcoming corner (a notable example is John Dupré 1987, 2012) – but it has flourished, which is an appropriate point to move on from philosophy in this world to religion in the next.

8

Religion

Religion: Warfare

Thanks particularly to two tendentious nineteenth-century accounts of the coming of evolution and its effects on the religious world – J. W. Draper's *History of the Conflict Between Religion and Science* (1875) and A. D. White's *History of the Warfare of Science with Theology in Christendom* (1896) – the popular metaphor for the relationship between science and religion after the *Origin* is that of "warfare" (Barbour 1997). Too well-known today are the biblical literalists, mainly American, self-named "Scientific Creationists," together with followers of the watered-down version "Intelligent Design Theory." They date back, before the *Origin*, to the early years of the nineteenth century. They like to characterize themselves as "traditional" Christians, but this is far from true. In the fourth century after Christ, St. Augustine (1991) had dealt, efficiently, with the matter of reading the Bible literally. With such a reading, at once one starts to run into trouble. Genesis tells us that light and dark were created on the First Day. It then assures us that we had to wait for the Fourth Day for the sun to make an appearance. Impossible! Undoubtedly influenced by his early, pre-Christian Manichaeism, a belief system that downgraded the importance of the Old Testament, Augustine argued that, although the Bible is true, through and through, sometimes it is necessary to interpret it allegorically. The Ancient Jews were not educated thinkers like fourth-century AD Romans. Descriptions too closely based on what God actually did, and they wouldn't have understood a word that was going on. God created, probably all at one time, and then explicated in such a way that people like Abraham and Isaac could catch the important truths.

In the American South, and even more so with those caught up with the country's push west, there were no Augustinians. That kind of theology made little headway with – was probably completely unknown to – such unsophisticated people (Noll 2002; Numbers 2006). These were not folk with an Ivy League education. In any case, they did not need cutting-edge theology. It was a harsh and unwelcoming country that was being settled: baking hot summers, consequent droughts, and winter storms, all matched by dangerous animals, not to mention the not always friendly humans long there and now being disturbed. Hurdles exacerbating the difficulties of having regular community life – stores, schools, churches, doctors, and hospitals. Folk like this needed simple, comforting rules of behavior. These were offered by itinerant preachers, fixed features, leading crusades thereby spreading the gospel. All importantly, there were new mechanical ways of printing. Ironically, in Scotland, this was the reason for Robert Chambers's economic success and hence his fervent belief in progress, the value-impregnated foundation of his evolutionary speculations; in the American South, this was the reason for freely available and affordable books, the value-impregnated foundation of biblical literalism. Given that most people were Protestant – *sola scriptura* – the Holy Book started to take a very prominent place: What should one believe? What should one do? What are the proper relationships between man and wife? How does one train and discipline one's children? Servants and slaves? What is owed to them? What is expected of them? Look to the Bible for advice (Numbers 1992).

The devil can cite Scripture for his purpose. So could those caught up in the controversy about slavery. Anti-slavery proselytizers, in the North, cited the Bible – the Beatitudes for instance – as evidence against the owning and subjection of other human beings. In the South, to the contrary, the Bible was cited in favor of slavery. Ephesians 6:5–9 was a particular favorite:

> 5 Slaves, obey your earthly masters with respect and fear, and with sincerity of heart, just as you would obey Christ. 6 Obey them not only to win their favor when their eye is on you, but as slaves of Christ, doing the will of God from your heart. 7 Serve wholeheartedly, as if you were serving the Lord, not people, 8 because you know that the Lord will reward each one for whatever good they do, whether they are slave or free.

St. Paul said this. Questioning is heresy. Civil War followed.

The North won, but with the collapse of Reconstruction and the end of hopes of catching up with modern civilization, the Bible, taken literally, continued to play a big role. In the South, in the newly occupied lands of the Midwest, and increasingly in poorer regions of large cities, where the already-existing inhabitants felt threatened by the influx of European immigrants: Catholic, with a leavening of Jews, Irish in Boston, Italians in New York, Polish in Chicago. The story of the Israelites in captivity to the Babylonians resonated with those defeated in the Civil War. On those whom He loves most, God lays the greatest burdens. Thinking like this was enshrined in a series of pamphlets, the *Fundamentals*, published at the beginning of the twentieth century (Marsden 1980). "The evidence for evolution, even in its milder form, does not begin to be as strong as that for the revelation of God in the Bible" (Wright 1910). Expectedly, given that the movement was caused as much by sociological factors as by anything theological, beliefs taken as obvious by Northerners, especially intellectuals and the educated, were anathema and despised. Sound familiar? MAGA. Darwinian evolution was at the top of this list, especially given the way in which it denied literalism. Warfare indeed.

The Problem of Evil

Not to conclude that warfare occurred only between literalism and modern science. To be honest, Darwin himself – his theory, that is – contributed to all of this. For a start, there is the problem of evil. Darwin worried about this one, writing just after the *Origin* to his American Presbyterian friend, the botanist Asa Gray:

> I had no intention to write atheistically. But I own that I cannot see, as plainly as others do, & as I shd wish to do, evidence of design & beneficence on all sides of us. There seems to me too much misery in the world. I cannot persuade myself that a beneficent & omnipotent God would have designedly created the Ichneumonidæ with the express intention of their feeding within the living bodies of caterpillars, or that a cat should play with mice. Not believing this, I see no necessity in the belief that the eye was expressly designed. (Letter to Asa Gray, May 22, 1860)

How can one believe in an all-powerful, all-loving God when there is so much pain and suffering in the world?

There are answers. Traditionally the problem is divided into two: natural evil – the Lisbon earthquake – and human-caused evil – the Final Solution. As far as natural evil is concerned, Richard Dawkins (1983) of all people has a powerful argument. With few exceptions – Descartes in his *Meditations* being one – it is agreed that God cannot do the impossible. God cannot make 2+2=5. To have functioning life, you must have a process that produces adaptation, as well as a livable environment. Processes such as Lamarckism or saltationism – evolution by lucky monsters – simply don't work. It is natural selection or nothing. But if natural selection, then you must let the world run according to unbroken law with its consequences – the struggle for existence, for example. You are bound to get pain and strife. Something such as childhood leukemia is just the luck of the draw. You produce complex machines as humans do and you are going to have failures, misfunctions. And to produce sunshine and fertile fields and so forth, you must pay the cost of earthquakes and the like.

What of moral evil? The usual God-exonerating argument is that it is a function of free will. Better to let people sin than to make them robots. If the latter, they do not have the potential to be good or bad, a consequence God most certainly did not want. Of course, you must now accept that the free will of Heinrich Himmler was equal to the worth of six million Jews. More modestly, of equal worth to one Anne Frank. But, as the saying goes, you cannot make an omelet without cracking eggs. We shall have more to say in Chapter 9 about the idea of free will. Here just rest with the fact that no believer is going to belittle the problem of evil. You might say that this is a problem with a solution beyond our ken. Invoking the distinction between faith and reason, one says simply that this is a place where you must leave things to faith. The point being made here is that, for those who insist on a solution backed by reason, there are traditional avenues to explore. For now, leave it at that.

Without Meaning?

As the years went by, particularly thanks to arguing with Asa Gray, as we see in the letter just quoted, Darwin became more and more inclined to think that his theory demolished the argument from design. He grew to agree with those who argued that it was the theist's God or no God at all. At the end of the large two-volume work he published later in 1868 – *The Variation of Animals and Plants under Domestication* – Darwin argued that he could not believe in a God who created all sorts of variations and, knowing the winner, stepped back and let nature run its course. Suppose "an architect were to rear a noble and commodious edifice, without the use of cut stone, by selecting from the fragments at the base of a precipice wedge-formed stones for his arches, elongated stones for his lintels, and flat stones for his roof" (249) (Figure 8.1). Would we truly want to say that the stones were random, given that God had been behind making them as they are? "The shape of the fragments of stone at the base of our precipice may be called accidental, but this is not strictly correct; for the shape of each depends on a long sequence of events, all obeying natural laws" (431). However, "An omniscient Creator must have foreseen every consequence which results from the laws imposed by Him. But can it be reasonably maintained that the Creator intentionally ordered, if we use the words in any ordinary sense, that certain fragments of rock should assume certain shapes so that the builder might erect his edifice?"

Natural selection destroys the argument from design, even as it keeps the major premise, that organisms are as-if designed. Thus, in his *Autobiography*

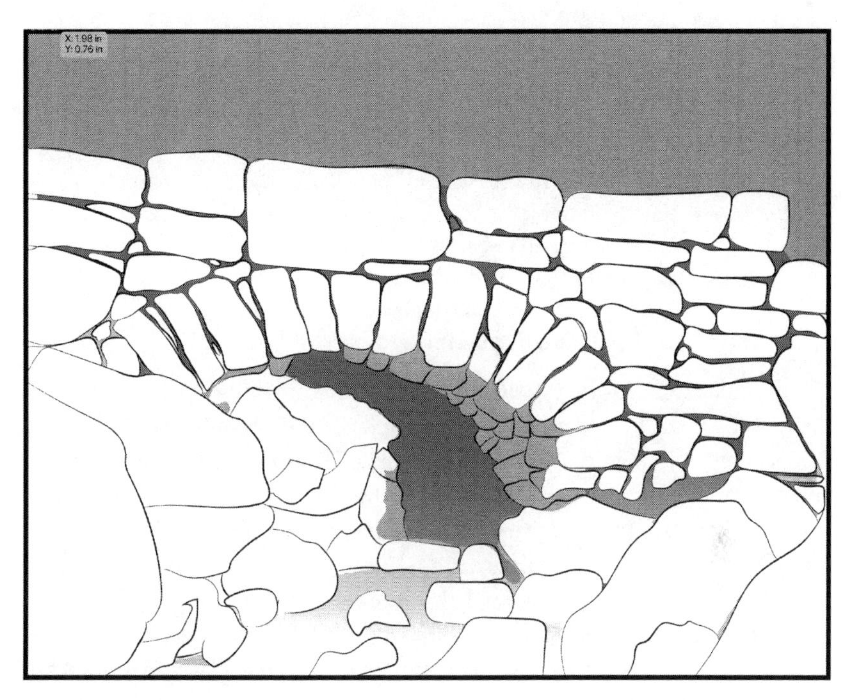

Figure 8.1 Stone bridge.

(written around 1875), Darwin felt able to write: "The old argument of design in nature, as given by Paley, which formerly seemed to me so conclusive, fails, now that the law of natural selection has been discovered." The Darwinian world is stripped of all meaning. Even more so if you consider the other side to meaning, not just individual adaptations but also the overall direction and purpose of evolution, the progressive advance up to humankind. Take this away and the very metaphysical structure of Christianity is destroyed. We are no longer made, uniquely, in the image of God.

By 1865, Thomas Hardy, poet and novelist, raised as a steady member of the Church of England, saw that a God of Darwin is simply indifferent to our needs and sufferings. It would be dreadful to have a God who hates us. Even worse, as is shown by his sonnet "Hap," to have a God indifferent to us:

> If but some vengeful god would call to me
> From up the sky, and laugh: "Thou suffering thing,
> Know that thy sorrow is my ecstasy,
> That thy love's loss is my hate's profiting!"

> Then would I bear it, clench myself, and die,
> Steeled by the sense of ire unmerited;
> Half-eased in that a Powerfuller than I
> Had willed and meted me the tears I shed.
>
> But not so. How arrives it joy lies slain,
> And why unblooms the best hope ever sown?
> – Crass Casualty obstructs the sun and rain,
> And dicing Time for gladness casts a moan ....
> These purblind Doomsters had as readily strown
> Blisses about my pilgrimage as pain.

All of this is simply terrifying. Catholics and Protestants made known their objections. St. George Mivart started as a Protestant, an enthusiast for Darwin's ideas, and a protégé of Huxley. He converted to Catholicism, and that was the beginning of the end. He turned into one of Darwin's most ferocious critics. On the Protestant side, an example of such frenetic writing comes from a vicar who spent most of his time cataloguing butterflies and birds: "If the whole of the English language could be condensed into one word, it would not suffice to express the utter contempt those invite who are so deluded as to be disciple of such an imposture as Darwinism" (Moore 1979, 197). This is not a man driven by reason alone.

Religion: Welcoming

It was to be expected, given his joking in the *Water-Babies*, that one ordained member of the Church of England who responded positively to the Darwinian ideas was Charles Kingsley. Thanking Darwin for a copy of the *Origin*, he wrote:

> I have gradually learnt to see that it is just as noble a conception of Deity, to believe that he created primal forms capable of self development into all forms needful pro tempore & pro loco, as to believe that He required a fresh act of intervention to supply the lacunas w^{h}. he himself had made. I question whether the former be not the loftier thought. (Letter to Darwin, November 18, 1859)

Darwin put an extract of this letter up front in later editions of the *Origin*.

Another enthusiast was the Reverend Baden Powell, Oxford Professor of Geology, father of the founder of the Scouting movement. From the publication of the *Origin* in 1859, before the publication of the *Origin* even, he was always empathetic to evolutionary ideas, defending them from charges of being anti-Christian (Powell 1855). *Essays and Reviews*, a highly controversial set of essays by liberal ("Broad-Church") Anglicans, six out of the seven being ordained, was published four months after the *Origin*. Baden Powell's contribution unambiguously endorsed Darwin's theory:

> A work has now appeared by a naturalist of the most acknowledged authority, Mr. Darwin's masterly volume on The Origin of Species by the law of natural selection, which now substantiates on undeniable grounds the very principle so long denounced by the first naturalists, the origination of new species by natural causes: a work which must soon bring about an entire revolution of opinion in favour of the grand principle of the self-evolving powers of nature. (Powell 1860, 139)

Expectedly, more conservative Anglicans, including the Archbishops of Canterbury and York, as well as our old friend, the Bishop of Oxford, Samuel Wilberforce, objected strenuously to the volume, and a charge of heresy was brought against two of the contributors. Eventually, the Privy Council exonerated the essayists: "hell was dismissed with costs, and took away from Orthodox members of the Church of England their last hope of eternal damnation" (Rowell 1974, 116).

Baden Powell started a tradition. Across the Atlantic, slavery opponent – his sister was Harriet Beecher Stowe, author of *Uncle Tom's Cabin* – congregationalist minister Henry Ward Beecher was dubbed the "most famous preacher in the nation." Having very publicly batted away an almost certainly justified charge of adultery, he turned his attention to evolution. As rhetorical in print as he was in the pulpit, he let flow:

> If single acts [of creation] would evince design, how much more a vast universe, that by inherent laws gradually builded itself, and then created its own plants and animals, a universe so adjusted that it left by the way the poorest things, and steadily wrought toward more complex, ingenious, and beautiful results! Who designed this mighty machine, created matter, gave to it its laws, and impressed upon it that tendency which has brought forth the almost infinite results on the globe, and wrought them into a perfect system? Design by wholesale is grander than design by retail. (Moore 1979, 221)

The more liberal clergy inclined to the grand metaphysical speculations of Herbert Spencer – after all, they themselves often tended to be in the business of grand metaphysical speculations – whereas the more traditional and conservative Christians often felt more comfortable with Darwin's thinking. His theory was based on solid Church of England natural theology – Paley and design, Malthus and struggle, for a start. In this vein, consider words penned towards the end of the nineteenth century by the High Anglican Oxford academic Aubrey Moore: "Darwinism appeared, and, under the guise of a foe, did the work of a friend." He continued that we must choose:

> Either God is everywhere present in nature, or He is nowhere. He cannot be here, and not there. He cannot delegate his power to demigods called "second causes." In nature everything must be His work or nothing. We must frankly return to the Christian view of direct Divine agency, the immanence of Divine power from end to end, the belief in a God in Whom not only we, but all things have their being, or we must banish him altogether. (Moore 1889, 73–74)

Unlike Hardy's indifferent deity, the God of Moore is always there, actively supporting the laws that make things tick. Including the story of the Sixth Day: the necessary end-appearance of humankind.

In this context, Moore was especially sensitive to the fact that Darwin did not ignore or deny teleology. Thinking in terms of final cause was central to his theory. It was rather that he gave a new understanding of teleology where the laws of nature could produce it. No need for special interventions. Moore was a Christian who felt no hesitation or shame in being a Darwinian. He was also a Christian who saw Darwin's worries about a God who put design into his laws as a plus rather than as a minus. The appearance of humankind was no chance event. "More than most who sought an accommodation with Darwinism, Moore was able to see clearly some positive theological gains. The occasional interventions of a remote deity could now be replaced by the constant sustaining and creative activity of the divine spirit. This has the significant benefit of correlating the Christian understanding of how God acts in nature and in history" (Fergusson 2009, 81). He is deeply involved in the stories of Israel and the coming of Jesus. He is not the God of the deists, who stands back and lets it all happen at a distance. There is value, value not of our choosing. It is the value given by God, not value existing independently in the world.

The Twentieth Century

Creationism, the Bible – especially Genesis – taken literally, moved smoothly into the twentieth century. However, it was not until after the First World War that things started to bubble over, when a schoolteacher in Tennessee, John Thomas Scopes, was prosecuted for teaching evolution (Larson 1997). Leading the attack was three-times presidential candidate William Jennings Bryan and the defence was by notorious secular thinker Clarence Darrow. The trial became a public spectacle, reported worldwide. Scopes was found guilty; although, on a technicality, his conviction was overturned on appeal. Predictably, the trial cast a dark shadow on what was permissible in the classroom. Although the curricula were determined state by state, decisions were ruled by the available textbooks. As always, sales figures rated more highly than dissemination of the truth. Publishers of biology textbooks, ever sensitive to the moods of the times, out did T. H. Huxley and his contemporaries in their eagerness to avoid the slightest reference to Darwin and his theory.

By the mid 1950s, people were becoming sensitive to the need for improved scientific education, especially in the light of Russian scientific achievements, notably the H-bomb (Rudolph 2015). Then, in 1957, Sputnik was put into space – another Russian success at the height of the Cold War. Now there was huge concern about the second-rate nature of contemporary American science, especially including American science education. New, up-to-date texts, culminating in *Biological Science: Molecules to Man* (1963), were commissioned and distributed by the Biological Sciences Curriculum Study Program. The evangelicals found that their children were being taught the hated evolution – as gospel, to use a phrase. Reaction was already in motion. In 1961, a biblical scholar and a hydraulic engineer, John C. Whitcomb and Henry M. Morris, published *Genesis Flood*, arguing against evolution and for the literal truth of the whole Holy Bible – six-day creation, universal deluge, parting of the Red Sea, and so on down to the present (Whitcomb and Morris 1961). It was hugely successful, not least because the emphasis on the Flood, rather than the Creation, reflected the fear of atomic conflict. The worst is yet to come: Armageddon.

Through the 1960s and 1970s, Creationism went from strength to strength. Highly effective were debates between Creationists, such as Morris and Gish, and somewhat naïve evolutionists who learnt to their dismay that, at circus-like events such as these, a good joke is far more effective than ten minutes of careful exposition of the Hardy–Weinberg law. Things came to a head in 1981, when the State of Arkansas passed a law insisting that, in biology classes of publicly funded schools of the state, Creationism (or Creation Science) be given "balanced treatment" along with the teaching of evolution (Ruse 1988b). After all, were they not incommensurable Kuhnian paradigms, neither absolutely true nor absolutely false (Ruse 2021a)? At once, the American Civil Liberties Union sprang into action, arguing that the law violates the First Amendment of the Constitution, because egregiously it mixes Church and State. Another trial ensued. It was realized that issues were going to be as much philosophical – what is the difference between science and religion? – as scientific or theological. Hence, along with the popular science writer Stephen Jay Gould and leading Protestant theologian Langdon Gilkey, I was an expert witness for the prosecution. The judge, William Overton, ruled firmly against Creationism. He said flatly:

> More precisely, the essential characteristics of science are:
>
> (1) It is guided by natural law;
> (2) It has to be explanatory by reference to natural law;

(3) It is testable against the empirical world;
(4) Its conclusions are tentative, i.e. are not necessarily the final word; and
(5) It is falsifiable. (Ruse and other science witnesses.)

Creation science … fails to meet these essential characteristics.

Evolutionary thinking is science. Creationism is religion and, as such, can have no place in the classrooms of state-funded schools.

Creationism had apparently met its final destiny. Yet, as always, things were never quite that easy. The Creationists quit trying to do things through legislation. Rather, very successfully pressuring individual school boards and the like, they worked at getting evolution-friendly textbooks banned from classes and from school libraries and all other places where young minds might encounter such heinous thinking. As importantly, as noted, they developed a smoother version of their position, Creationism-lite, which they called Intelligent Design Theory. Now the emphasis was less on the Bible taken literally and more on the need of an intelligence – more precisely, Intelligence – to get all started and going (Johnson 1991; Behe 1996; Dembski 1998). Smoother, but no more convincing. Intelligent Design Theory also met its destiny, this time in 2005, in the Kitzmiller v. Dover [Pennsylvania] Area School District trial. It was ruled that intelligent design was not science, that it "cannot uncouple itself from its creationist, and thus religious, antecedents," and that hence teaching it in public schools violated the Establishment Clause of the First Amendment to the United States Constitution.

Reason versus Faith

One would be naïve to think that this, or ever will be, the end of the story. Apparently, in the state of Louisiana, Creationism is still taught openly in public schools. One of the chief legal enablers is the present speaker of the House of Representatives, Mike Johnson (Blumenthal 2023). The man second in line to the Presidency of the United States of America uses his time and talents to ensure that the schoolchildren of his state, Louisiana, are taught that the six-day creation story of Genesis 1 is literally true. St. Augustine must be revolving in his grave. For now, if only for the sake of discussion, let us pretend that Louisiana does not exist. Rule out Creationism, both the traditional version as found in *Genesis Flood* and the more recent, smoother version – Intelligent Design Theory. Where does Darwin's theory leave more conventional versions of Christianity, Catholic and/or Protestant? Go back to the brief discussion in Chapter 1. A distinction is always drawn between reason and faith. Trying to understand God, and our relationship to Him, might be simply a matter of reason. What grounds do logic and empirical experience give for insights into the deity? This area of inquiry, known as "natural theology," encompasses the arguments for God's existence, as well as challenges the problem of evil. Without saying that Darwin's theory unassisted destroys the whole enterprise – arguments and conclusions – we have seen that it does make some of the traditional arguments rather less plausible than before. The argument from design no longer has us in its inescapable grip. In the opinion of Richard Dawkins (1986), Darwinism makes it possible to be an "intellectually fulfilled atheist." This does not mean that one must be an atheist or that the Design argument no longer has any force or value. But the relationship is changed. In the memorable words of Britain's greatest theologian of the nineteenth century, John Henry Newman: "I believe in design because I believe in God; I do not believe in God because I believe in design" (Newman 1973, 97).

None of this is that troubling to the believer. It has always been the case that Christianity prioritizes faith over reason. We (Plantinga is emphatic about this) have as it were a hotline to God, what Jean Calvin called a *sensus divinitatis*:

> That there exists in the human minds and indeed by natural instinct, some sense of Deity, we hold to be beyond dispute, since God himself, to prevent any man from pretending ignorance, has endued all men with some idea of his Godhead, the memory of which he constantly renews and occasionally enlarges, that all to a man being aware that there is a God, and that he is their Maker, may be condemned by their own conscience when they neither worship him nor consecrate their lives to his service. (Calvin 1536, 1, 3, 1)

The high status the Christian accords to faith, as we know, starts with Jesus and the disciple "Doubting" Thomas, who refuses without proof to acknowledge Jesus as the risen lord. "Jesus saith unto him, Thomas, because thou hast seen me, thou hast believed: blessed are they that have not seen, and yet have believed" (John 20:27–29). Faith over reason, a position affirmed and reaffirmed down to the present. Well known is a strain of Christian thought – most earnestly argued for by the Danish theologian Søren Kierkegaard – that affirms that were natural theology successful, this would destroy revealed theology. The very point of faith is that it demands a leap into the absurd. If we could just prove events such as the Resurrection to be historically authentic, then there would be no merit in accepting it on faith.

In short, whatever Darwinism does to natural theology, it leaves revealed theology or religion untouched. Nonetheless, there are obviously those Darwinians who think the very idea of faith as a guide to truth is ludicrous. Jerry Coyne writes:

> My claim is this: science and religion are incompatible because they have different methods for getting knowledge about reality, have different ways of assessing the reliability of that knowledge, and, in the end, arrive at conflicting conclusions about the universe. "Knowledge" acquired by religion is at odds not only with scientific knowledge, but also with knowledge professed by other religions. In the end, religion's methods, unlike those of science, are useless for understanding reality. (Coyne 2015, xvi)

Whether or not this be true, the supposed incompatibility of science and faith is not based on Darwin's theory as such. It is based rather on the assumption of the materialistic nature of science, which simply dismisses faith in psychological terms such as fear of the unknown – the feeling expressed in Hardy's poem "Hap" that everything is, in the language of Albert Camus, absurd. The important concluding thought is that, however faith turns out to be, there is nothing we have seen thus far making an appeal to this nonscientific grasp of reality impossible or even unreasonable. Darwinism implies much. It does not imply everything.

Organicism

Thus far, as Coyne makes clear, we have been working very much within the machine root metaphor. Has the organism root metaphor nothing to say about the Darwinism–religion relationship? To be honest, so long as we stay focused on natural selection, not a great deal. Woodger, the organicist, focused right in on selection, wanting to take it from the discussion. However, past argument should key us to think that evolution as such – which is not constrained by the mechanistic demands of selection – might have a contribution to make – on empirical grounds, if none other (Seeman, Dubin, and Seeman 2003). For a start, unlike the machine-metaphor picture, the organism-metaphor picture is value impregnated, as also is the Christian picture. The organicist sees the world as having value in itself. The Christian does not see the world as having value in itself – all value comes from God – but as with the organic metaphor, the way we see the world is value impregnated. Remember: "And God saw every thing that he had made, and, behold, it was very good. And the evening and the morning were the sixth day" (Genesis 1:31).

Additionally, the organic metaphor sees progress in evolution – blob to human. This is precisely the position of the Christian:

> So God created mankind in his own image,
> in the image of God he created them;
> male and female he created them.
>
> (Genesis 1:27)

Promising. Do we actually see the organic metaphor at work? Turn to the thinking of the author of *L'évolution créatrice* (1907; translation, *Creative Evolution* 1911), the French "vitalist" philosopher Henri Bergson. He posited a kind of Aristotelian life force, the *élan vital*, to explain the upward course of evolution. Bergson was not too keen on a teleological system as such. It seems to have a kind of predetermined finality about it. The acorn grows into the oak tree – it attains its end without need of input on the way. No other option.

> The doctrine of teleology, in its extreme form, as we find it in Leibniz for example, implies that things and beings merely realize a programme previously arranged. But if there is nothing unforeseen, no invention or creation in the universe, time is useless again. As in the mechanistic hypothesis, here again it is supposed that all is given. Finalism thus understood is only inverted mechanism. (Bergson 1911, 39)

Although influenced by Spencer, it is the Englishman being criticized here. Bergson thought Spencer too much of a mechanist. Nevertheless, Bergson does think there is something end-directed about everything. It is here that, showing the influence of Schelling, the *élan vital* plays its crucial role. It is this that lies behind evolution:

> An original impetus of life, passing from one generation of germs to the following generation of germs through the developed organisms which bridge the interval between the generations. This impetus, sustained right along the lines of evolution among which it gets divided, is the fundamental cause of variations, at least of those that are regularly passed on, that accumulate and create new species. (87)

It leads to a creative process:

> If now we are asked why and how it is implied therein, we reply that life is, more than anything else, a tendency to act on inert matter. The direction of this action is not predetermined; hence the unforeseeable variety of forms which life, in evolving, sows along its path. But this action always presents, to some extent, the character of contingency; it implies at least a rudiment of choice. (96)

It is still inward forces driving change. Most particularly, we get the evolution of humans. "So that, in the last analysis, man might be considered the reason for the existence of the entire organization of life on our planet" (185).

We are certainly not getting a familiar picture of Christianity, even though, despite his remaining a Jew lest he seem to be abandoning his coreligionists under the burdens of Vichy France, Bergson felt he was sufficiently close that he could ask the Catholic Church for prayers after his death. Most immediately, we think of the vital impetus, the *élan vital*, as in some sense representative of the Divine. It is not material; it works on – and hence is separate from – our physical world. Above all, it is creative – it has free choice – and that is exercised in creating the truly superior, truly unique species: humankind, the apotheosis of creative evolution. There are still lots of questions, but we are edging closer to a more familiar theological world picture – "Let us make man in our image, after our likeness" – a conclusion confirmed by the reply Bergson penned to a (Catholic priest) critic in 1912. He stressed that for him God was the source of all, independent of His creation, which was done by Him freely. Hence, the material is dependent on God, not God on the material. He added that he is committed to "the idea of a God, creator and free, the generator of both Matter and Life, whose work of creation is continued on the side of Life by the evolution of species and the building up of human personalities. From all this emerges a refutation of monism and of pantheism" (Gunn 1920, 129).

This in the realm of religion. Bergson was not without his supporters in the scientific world. It will come as no shock to learn that one immediate enthusiast was Julian Huxley, already on his way to his non-Darwinian vision of the evolutionary process. He concluded his first book, *The Individual in the Animal Kingdom* (1912, vii), by saying: "It will easily be seen how much I owe to M. Bergson, who, whether one agrees or no with his views, has given a stimulus (most valuable gift of all) to Biology and Philosophy alike." (Huxley was French-speaking so he could work from the original edition, of 1907.) It was, however, in religion where Bergson had his greatest – certainly most notorious – effect, for all that his most famous disciple claimed he was still in the realm of science.

Omega Point

> If this book is to be properly understood, it must be read not as a work on metaphysics, still less as a sort of theological essay, but purely and simply as a scientific treatise. The title itself indicates that. This book deals with man solely as a phenomenon; but it also deals with the whole phenomenon of man. (Teilhard de Chardin 1955, 28)

Those are the opening lines of the *Phenomenon of Man* by Father Pierre Teilhard de Chardin SJ. If ever they offer a prize for the most misleading opening of any book, it will be a heavy favorite. Teilhard de Chardin's book may be many things. Purely and simply a work of science it is not, for all that the eminent (secular) historian Jean Gayon (2013) tells us that Teilhard was certainly the most brilliant French paleontologist of the first part of the twentieth century. More accurately, it offered the most systematic and audacious attempt at integration of evolutionary theory and the Christian religion.

The basic thesis of the *Phenomenon of Man*, published posthumously in 1955, is quite simple. The world and life within it reveal a process of evolution of ever-greater complexity, taking us from the inorganic through the simplest organisms, up through various stages of existence to the highest, the "noösphere." This is the world of humankind, culminating in something Teilhard called the "Omega Point": "Our picture is of mankind labouring under the impulsion of an obscure instinct, so as to break out through its narrow point of emergence and submerge the earth; of thought becoming number so as to conquer all habitable space, taking precedence over all other forms of life; of mind, in other words, deploying and convoluting the layers of the noösphere" (Teilhard de Chardin 1955, 190). Very controversially, Teilhard identified the climax, the Omega Point, with God as incarnated in Jesus Christ: "The universe fulfilling itself in a synthesis of centres in perfect conformity with the laws of union. God, the Centre of centres. In that final vision the Christian dogma culminates" (293). A final vision perfectly coinciding with the Omega Point.

Inventive, but, overall, not entirely unfamiliar. The vision of evolution that is the backbone of the *Phenomenon of Man* is Bergsonian through and through – above all, organicist, in seeing ever greater complexity, driving towards some higher end: humans! Little surprise that Julian Huxley wrote an enthusiastic introduction to the English translation of the *Phenomenon of Man*. Perhaps more surprise to learn that Theodosius Dobzhansky was president of the American Teilhard Association! At the end of a popular book on human evolution – *Mankind Evolving* (1961) – Dobzhansky wrote: "To modern man so spiritually embattled in this vast and ostensibly meaningless universe, Teilhard de Chardin's evolutionary idea comes as a ray of hope" (348). In a letter to John C. Greene, historian of science, he admitted to being a panentheist (Greene and Ruse 1996). Dobzhansky, a practicing Christian (Russian Orthodox), seems to have been playing the same game as Simpson (who wrote an enthusiastic blurb for Dobzhansky's book): progress and values in popular writings (1949), no progress or any nonepistemic values in professional science (1944, 1953). Organicist in the popular world; mechanist in the professional world. Julian Huxley, by contrast, less a professional biologist and more a man of science – head of the London Zoo, first Director-General of UNESCO, seemingly always on the radio (later television) – had no such qualms about introducing values into his vision of evolution. Organicist all the way. Simpson's willingness to endorse Dobzhansky may have been a function of his close friendship with Teilhard, understandable considering Gayon's assessment of Teilhard as a scientist; although, as he told many people (for instance, Steven J. Gould, April 14, 1976), Simpson thought Teilhard a hypocrite, a supposedly celibate Jesuit, travelling openly with his mistress (Ruse 1996, 427).

If you view Teilhard through the lens of a Darwinian mechanist, you may have science on your side, but you miss what he is about. Does Teilhard then have no worries about running together what others might consider the separate domains of science and religion? Apparently not. What does interest Teilhard is values. As a through and through organicist, he sees ever greater value the closer we get to humankind, with the corollary that the more humankind grows in complexity, the more value we have and the more value we want to generate: "conquered by the sense of the earth and human sense, hatred and internecine struggles will have disappeared in the ever warmer radiance of Omega. Some sort of unanimity will reign over the entire mass of the noösphere. The final convergence will take place in peace" (287).

We come to purpose and ends, that which gives the full (and only) meaning to Teilhard's world system. With the noösphere, taken out of context, we are still in the domain of the secular organicist; but then, with the Omega Point, we go beyond. All is put in a Christian context: "The end of the world: the overthrow of equilibrium, detaching the mind, fulfilled at last, from its material matrix, so that it will henceforth rest with all its weight on God-Omega. The end of the world: critical point simultaneously of emergence and emersion, of maturation and escape" (287). Taken all in all, a wonderful world vision, even if it must be said that *C'est magnifique, mais ce n'est pas la science.*

Process Theology

In 1924, the Harvard faculty was enriched by the arrival of the English logician Alfred North Whitehead – who, with Bertrand Russell, was deservedly famous for the attempt (in their magnum opus *Principia Mathematica* 1913) to show that mathematics follows deductively from the laws of logic. People joked that the first time Whitehead gave a lecture at Harvard was the first time he had ever been in a philosophy class. This did not deter him from moving into metaphysics, and giving a series of lectures, published as *Science and the Modern World* (Whitehead 1926). Openly declaring himself an organicist – don't forget, he had worked with Russell – Whitehead called for "the abandonment of the traditional scientific materialism, and the substitution of an alternative doctrine of organism" (99). Continuing: "Nature exhibits itself as exemplifying a philosophy of the evolution of organisms subject to determinate conditions" (115). Although it seems likely that Whitehead got this secondhand, there is no mystery about the major influence here (Gare

2002): "Nature should be Mind made visible, Mind the invisible nature. Here then, in the absolute identity of Mind in us and Nature outside us, the problem of the possibility of a Nature external to us must be resolved" (Schelling 1797, 42). Consolidating his ideas, Whitehead writes: "The final summary can only be expressed in terms of a group of antitheses, whose apparent self-contradictions depend on neglect of the diverse categories of existence. In each antithesis there is a shift of meaning which converts the opposition into a contrast" (Whitehead 1929, 347).

Thus far, in our discussion of organicism, the Augustinian God goes unscathed. His creation may be organic. He is not. That Henri Bergson should seriously have contemplated joining the Catholic Church suggests strongly that he was not about to rock the boat in that direction. Likewise, Teilhard's God, for instance, is very recognizable. Everything else may be in motion. He is not. Turn now to Alfred North Whitehead. Becoming the leading, and highly influential, exponent of the "philosophy of organism," he did not just discard the traditional God of Augustine, he was repelled in horror. Undoubtedly influenced by the tragedy of the Great War – in which he and his wife lost their son Eric, aged nineteen, in the Royal Flying Corps – Whitehead turned his back on such an appallingly cruel creature. His God was down there in the trenches, with us trembling with fear while we awaited the order to climb the ladder and face the enemy's bullets as we scrambled across no man's land. His God was with a young lad as his plane rushed to the earth, carrying him from the optimistic joy of a life just beginning to certain death. This is the God of the poet Pattiann Rogers:

> And maybe he suffers from the suffering
> Inherent to the transitory, feeling grief himself
> For the grief of shattered beaches, disembodied bones
> And claws, twisted squid, piles of ripped and tangled,
> Uprooted turtles and crowd rock crabs and Jonah crabs,
> Sand bugs, seaweed and kelp.
>
> (Rogers 2001, 182–83)

Whitehead and his followers (who, hardly surprisingly, included a very enthusiastic J. H. Woodger) washed their hands of a God who feels no compassion for the family when they learn that their child has leukemia, incurable and with pain that no human should have to face. Nothing to do with the God who is unmoved – cannot be moved – because He is eternal and unchanging. They wanted a God in the world rather than a God who is in some sense logically separate. "Nature should be Mind made visible, Mind the invisible nature." Whitehead writes:

> The vicious separation of the flux from the permanence leads to the concept of an entirely static God, with eminent reality, in relation to an entirely fluent world, with deficient reality. But if the opposites, static and fluent, have once been so explained as separately to characterize diverse actualities, the interplay between the thing which is static and the things which are fluent involves contradiction at every step in its explanation. (Whitehead 1929, 346)

Concluding: "The final summary can only be expressed in terms of a group of antitheses, whose apparent self-contradictions depend on neglect of the diverse categories of existence. In each antithesis there is a shift of meaning which converts the opposition into a contrast." Thus:

> It is as true to say that the World is immanent in God, as that God is immanent in the World.
>
> It is as true to say that God transcends the World, as that the World transcends God.
>
> It is as true to say that God creates the World, as that the World creates God. (347–48)

That means that God Himself must be part of the evolutionary process. Not the world itself – pantheism – but in all parts of the world. "Panentheism"! And since the process of evolution involves struggle (the Englishman Whitehead is part of the Malthusian culture that produced Darwin), this means that God too must be suffering and striving: "he suffers from the suffering Inherent to the transitory." He has, as the story of Jesus shows full well, emptied Himself of His powers and works alongside us. "He made himself nothing by taking the very nature of a servant, being made in human likeness" (Philippians 2:7). This is known as "kenosis." As we have just seen, for many this is a very radical break with traditional Christian theology. There, God is said to be "impassible." Anselm: "For when thou beholdest us in our wretchedness, we experience the effect of compassion, but thou dost not experience the feeling" (1903, 13). Aquinas: "To sorrow, therefore, over the misery of others does not belong to God" (1952, I, 21, 3). The God of Whitehead would be in Bergen-Belsen, lying and suffering with Anne Frank as she died from typhoid – "delirious, terrible, burning up." The God of Whitehead would also be the God who conquered, who made available her diary, which has been and continues to be such a huge inspiration to her fellow human beings. Especially to her fellow young people. "In spite of everything I still believe that people are really good at heart. I simply can't build up my hopes on a foundation consisting of confusion, misery, and death" (July 15, 1944).

Postscript

Conclude here. We started this discussion with the "warfare" thesis. It cannot be denied that this is certainly not unknown in the natural-selection/religion interface, rather less so in the evolution/religion interface. Today, together with the Creationists, it is the New Atheists who keep the battlefield active (LeDrew 2016). The uniting theme is a detestation of religion, all religion. Richard Dawkins is eloquent: "The God of the Old Testament is arguably the most unpleasant character in all fiction: jealous and proud of it; a petty, unjust, unforgiving control-freak; a vindictive, bloodthirsty ethnic cleanser; a misogynistic, homophobic, racist, infanticidal, genocidal, filicidal, pestilential, megalomaniacal, sadomasochistic, capriciously malevolent bully" (Dawkins 2006, 31). Dawkins is "inclined to follow" the author of *Zen and the Art of Motorcycle Maintenance* (Robert Pirsig) when he says: "When one person suffers from a delusion, it is called insanity. When many people suffer from a delusion, it is called religion" (Dawkins 2006, 5).

There is more, much more, in this vein. However, while the New Atheists might make the most noise, they are not the only voices. As we have seen in this chapter, there have been and still are many other voices, with very different conclusions. For some, indeed, the coming of evolution, not excluding the coming of Darwinism, is one of the freshest and most exciting events in Christian theological history. Leave things at that.

9

Literature

Early Days

We can say with some confidence that, in Britain, as soon as the *Origin* was published, regular people (that is, middle-class people who were not particularly knowledgeable about science) knew not only about evolution but also about natural selection (Ruse 2017a). Charles Dickens – who had responded favorably to *Vestiges* – was immediately sensitive to the *Origin*, having Pip, the hero of *Great Expectations*, say of his five dead siblings that they "gave up trying to get a living, exceedingly early in that universal struggle" ([1860] 1948, 1). Just an aside, but it lets us infer that the Malthus–Darwin theme is right up there in popular consciousness. Dickens himself would have known all about Darwin's theory because in the weekly magazine he edited, *All the Year Round* (circulation about 100,000) – *Great Expectations* was published in serial form starting in December 1860 – he carried two articles in mid 1860 and another early in 1861 that discussed the *Origin* and natural selection carefully and sympathetically, even favorably (Anon. 1860a, b, 1861):

> How, asks Mr. Darwin, ... have all these exquisite adaptations of one part of the organisation to another part, and to the conditions of life, and of one distinct organic being to another, been perfected? He answers, they are so perfected by what he terms Natural Selection – the better chance which a better organised creature has of surviving its fellows – so termed in order to mark its relation to Man's power of selection. Man, by selection in the breeds of his domestic animals and the seedlings of his horticultural productions, can certainly effect great results, and can adapt organic beings to his own uses, through the accumulation of slight but useful variations given to him by the hand of Nature. But Natural Selection is a power incessantly ready for action, and is as immeasurably superior to man's feeble efforts, as the works of Nature are to those of Art. Natural Selection, therefore, according

> to Mr. Darwin – not independent creations – is the method through which the Author of Nature has elaborated the providential fitness of His works to themselves and to all surrounding circumstances. (Anon. 1860a)

The author, David Thomas Ansted (1814–80), was a professional geologist and long-time acquaintance of Darwin. The status of Darwin as a scientist and his worth as a human being are likewise strongly underlined. Although the pieces are ostensibly agnostic about Darwin's work, opponents are referred to as "timid" and, overall, the sentiment (about something explicitly labeled "mechanical") is very positive:

> We are no longer to look at an organic being as a savage looks at a ship – as at something wholly beyond his comprehension; we are to regard every production of nature as one which has had a history; we are to contemplate every complex structure and instinct as the summing up of many contrivances, each useful to the possessor, nearly in the same way as when we look at any great mechanical invention as the summing up of the labour, the experience, the reason, and even the blunders, of numerous workmen. (1860b, 299)

One doubts that Dickens was amazed by any of this for it was not a one-off by the author. In a book Ansted published later in 1860, *Geological Gossip*, the treatment is even more favorable. Just as well. Ansted was literally indebted to Darwin, who a year or two later forgave several hundred pounds owed to him (Ruse 2017a, 62). One doubts St. George Mivart would have been so fortunate, although he would hardly have needed help since his family owned Claridge's Hotel.

What is significant is that Darwin's ideas, discussed favorably by the middle of 1860, would have been very widely disseminated. A hundred thousand copies probably means that up to half a million people learnt about Darwin. In those pretelevision days, it was customary for families to get together and for one member of the group – often the wife/mother – to read aloud from such magazines as that of Dickens. Emma read to the Darwin family every evening. This was not just mind-improvement. It was in magazines such as that of Dickens where one got the novels serialized. Just like watching episodes of a television series.

Darwinism and Fiction

Creative thinker after creative thinker wrestled with the meaning and consequences of Darwin's thinking. Natural selection was a recurrent theme. Novelist George Gissing's *New Grub Street*, published about ten years after

Darwin's death, is a brilliant exemplar of how Darwinian theory can underly and structure a novel. Both sides gain. You have a rattling good read and at the same time a terrific advertisement for the theory, not just as esoteric science but as something commonsensical explaining and giving meaning to everyday life. Gissing tells the story of two young men – Edwin Reardon and Jasper Milvain – trying to make their ways in the London literary world in around 1880. Edwin, highly talented, refuses to compromise with vulgar taste. Jasper – well, let him speak for himself:

> You have no faith. But just understand the difference between a man like Reardon and a man like me. He is the old type of unpractical artist; I am the literary man of 1882. He won't make concessions, or rather, he can't make them; he can't supply the market. I – well, you may say that at present I do nothing; but that's a great mistake, I am learning my business. Literature nowadays is a trade. Putting aside men of genius, who may succeed by mere cosmic force, your successful man of letters is your skillful tradesman. He thinks first and foremost of the markets; when one kind of goods begins to go off slackly, he is ready with something new and appetising. He knows perfectly all the possible sources of income. (Gissing [1891] 1976, 38–39)

Edwin dies – a somewhat common occurrence in Victorian novels – and Jasper gets his widow and the editorship of a prized journal. Edwin is forgotten. "Amy sprang up and threw her arms about her husband's neck, uttering a cry of delight." Jasper is appreciative. "I owe my fortune to you, dear girl. Now the way is smooth!" (549).

Gissing's genius is to challenge (and change) the reader's thinking. Initial reaction is that, with respect to Darwinism – the survival of the fittest – what happens is not very positive. After all, the basic assumption is that Edwin is significantly fitter than Jasper. This is to miss the real point and fail to see that the author is a far truer Darwinian. Absolute merit, epistemologically or ethically, is not the ultimate measure in the Darwinian world. Gissing knew that Darwin had stressed that the struggle is not necessarily one competing organism directly against another but rather with the environment. "Two canine animals in a time of dearth, may be truly said to struggle with each other which shall get food and live. But a plant on the edge of a desert is said to struggle for life against the drought, though more properly it should be said to be dependent on the moisture" (Darwin 1859, 62). One plant is being selected rather than another, but because it does better in the environment and not because it beats out the other directly.

It is the same story with our two writers. Gissing knew full well that in many respects Edwin is better than Jasper – certainly when it comes to literary talent. But – and this is the Darwinian backing to the novel – success is not a matter of climbing some absolute chain of perfection. There are hints that success is amoral and that in the Darwinian world it is all a matter of being tougher and pushier than the competitor, for people as for books: "Speaking seriously, we know that a really good book will more likely than not receive fair treatment from two or three reviewers; yes, but also more likely than not it will be swamped in the flood of literature that pours forth week after week, and won't have attention fixed long enough upon it to establish its repute" (Gissing 1891). The Darwinian implications? For once, the cultural evolutionary analogy pays off.

> The struggle for existence among books is nowadays as severe as among men. If a writer has friends connected with the press, it is the plain duty of those friends to do their utmost to help him. What matter if they exaggerate, or even lie? The simple, sober truth has no chance whatever of being listened to, and it's only by volume of shouting that the ear of the public is held. (Gissing 1891, 493)

There is more to the story than that. Jasper may not be as gifted as Edwin, but – crucially Darwinian – he has a vitality that Edwin does not have. More than some of his girlfriends too.

> "You hear?"
>
> Marian had just caught the far-off sound of the train. She looked eagerly, and in a few moments saw it approaching. The front of the engine blackened nearer and nearer, coming on with dread force and speed. A blinding rush, and there burst against the bridge a great volley of sunlit steam. Milvain and his companion ran to the opposite parapet, but already the whole train had emerged, and in a few seconds it had disappeared round a sharp curve. The leafy branches that grew out over the line swayed violently backwards and forwards in the perturbed air.
>
> "If I were ten years younger," said Jasper, laughing, "I should say that was jolly! It enspirits me. It makes me feel eager to go back and plunge into the fight again."
>
> "Upon me it has just the opposite effect," fell from Marian, in very low tones. (63)

As we know, Jasper was to find marital happiness elsewhere. And in case the reader does not get the point, we learn that Jasper's wife – the widow of Edwin – was not living her life blindly, without direction: "Though she could not undertake the volumes of Herbert Spencer, she was intelligently acquainted with the tenor of their contents; and though she had never opened one of Darwin's books, her knowledge of his main theories and illustrations was respectable" (397). Need one say more?

Figure 9.1 Eugene and Lizzie on their wedding day.

When it came to storytelling, sexual selection was even more of a guide than natural selection. Dickens led the way in his last finished novel, *Our Mutual Friend* (1865). A major theme is a battle – male combat – for the hand of the waterman's daughter, Lizzie Hexam, between an indolent lawyer (barrister), Eugene Wrayburn – "'And I,' said Eugene, 'have been "called" seven years, and have had no business at all, and never shall have any. And if I had, I shouldn't know how to do it'" (Dickens 1865, 19) – and an overly sensitive schoolmaster – "Bradley Headstone, in his decent black coat and waistcoat, and decent white shirt, and decent formal black tie, and decent pantaloons of pepper and salt, with his decent silver watch in his pocket and its decent hair-guard round his neck, looked a thoroughly decent young man of six-and-twenty" (Dickens 1865, 178–79). Wrayburn wins the prize in the end (Figure 9.1), but only by surviving a murderous attack on him by Headstone, from which he is rescued by Lizzie. What makes the story powerfully compelling is that the antagonists are so very different, offering a distinctive choice. Wrayburn, embedded in his upper-middle-class status to an extent that he refuses to exert himself to the full and realize his potential, and Headstone, insecure in his lower-middle-class status, working nonstop to improve his place in the world and yet fearing – knowing – that this will be all to no avail.

Sexual selection is working flat out, making Wrayburn by far the preferable option – female choice – but Lizzie, from the working class (if that, given her disreputable father), is unable to take this option because of the class difference between them. She loathes Headstone: "'There is a certain man,' said Lizzie, 'a passionate and angry man, who says he loves me, and who I must believe does love me. He is the friend of my brother. I shrank from him within myself when my brother first brought him to me; but the last time I saw him he terrified me more than I can say.'" She adores Wrayburn:

> "It was late upon a wretched night," said Lizzie, "when his eyes first looked at me in my old river-side home, very different from this. His eyes may never look at me again. I would rather that they never did; I hope that they never may. But I would not have the light of them taken out of my life, for anything my life can give me."

But it cannot be: "Yes, he's a gentleman. Not of our sort; is he?" Until, through the evil actions of Headstone and the noble actions of Lizzie, Wrayburn is made so very vulnerable that the class divisions can be overcome:

> "Undraw the curtains, my dear girl," said Eugene, after a while, "and let us see our wedding-day."
>
> The sun was rising, and his first rays struck into the room, as she came back, and put her lips to his. "I bless the day!" said Eugene. "I bless the day!" said Lizzie.
>
> "You have made a poor marriage of it, my sweet wife," said Eugene. "A shattered graceless fellow, stretched at his length here, and next to nothing for you when you are a young widow."
>
> "I have made the marriage that I would have given all the world to dare to hope for," she replied. (752–53)

Darwinism and Poetry

We have seen already that poets picked up on evolutionary themes. Notably, remember Tennyson who used the message of *Vestiges* to find hope and reason for his friend's early death.

> Whereof the man, that with me trod
> This planet, was a noble type
> Appearing ere the times were ripe,
> That friend of mine who lives in God.

After the *Origin*, the tendency was more to find in Darwinism a reason for a loss of faith. We know that, although he did not publish until later, by 1866 Thomas Hardy was already finding reason to fear that in the new Darwinian world all is meaningless. In New England, Emily Dickinson (1960) likewise wrestled with the God question and, in an age of Darwin, the problem of indifference; although, at times one gets the sense that Darwin is being used more to back up a degree of non-belief that she had acquired already.

> It's easy to invent a Life—
> God does it—every Day—
> Creation—but the Gambol
> Of His Authority—
> It's easy to efface it—
> The thrifty Deity
> Could scarce afford Eternity
> To Spontaneity—
> The Perished Patterns murmur—
> But His Perturbless Plan
> Proceed—inserting Here—a Sun—
> There–leaving out a Man—

Emily Dickinson's God seems almost casual and indifferent to suffering. Far from being all-loving, He is a source of the troubled nature of our world:

> Apparently with no surprise
> To any happy Flower
> The Frost beheads it at its play—
> In accidental power—
> The blonde Assassin passes on—
> The Sun proceeds unmoved
> To measure off another Day
> For an Approving God—

"Blonde Assassin"? "Approving God"? Not exactly an enthusiastic endorsement for the existence of the God of the New Testament! What we do not get is Hardy's despair at a God who is indifferent, but rather contempt for a God who loads us down with guilt:

> "Heavenly Father"—take to thee
> The supreme iniquity
> Fashioned by thy candid Hand
> In a moment contraband—

Though to trust us—seem to us
More respectful—"We are Dust"—
We apologize to thee
For thine own Duplicity

We are a long way from:

My richest gain I count but loss
And pour contempt on all my pride.

(Watts 1707)

In the Darwinian world, if any blame is to be accorded, God is ahead of us in the queue (Brantley 2014).

Finally, in a lighter vein, there is the poetry of Constance Naden. Her delightful poem "Natural Selection" – truly more about sexual selection – was written in around 1885.

I HAD found out a gift for my fair,
I had found where the cave men were laid:
Skulls, femur and pelvis were there,
And spears that of silex they made.

But he ne'er could be true, she averred,
Who would dig up an ancestor's grave—
And I loved her the more when I heard
Such foolish regard for the cave.

My shelves they are furnished with stones,
All sorted and labeled with care;
And a splendid collection of bones,
Each one of them ancient and rare;

One would think she might like to retire
To my study—she calls it a "hole"!
Not a fossil I heard her admire
But I begged it, or borrowed, or stole.

But there comes an idealess lad,
With a strut and a stare and a smirk;
And I watch, scientific, though sad,
The Law of Selection at work.

Of Science he had not a trace,
He seeks not the How and the Why,
But he sings with an amateur's grace,
And he dances much better than I.

And we know the more dandified males
By dance and by song win their wives—
'Tis a law that with avis prevails,
And ever in Homo survives.

Shall I rage as they whirl in the valse?
Shall I sneer as they carol and coo?
Ah no! for since Chloe is false
I'm certain that Darwin is true.

(Naden 1999, 207–8)

It is worth noting that, at this time, women were generally most unwelcome in the realm of professional science. Huxley, for instance, would not allow women into his university science classes. To Lyell, who wanted to let women into the Geological Society, taking note of the enlightened way in which his own daughters had been raised, Huxley wrote: "You know as well as I do, that other people won't do the like, and five sixth of women will stop in the doll stage of evolution, to be the stronghold of parsondom, the drag on civilization, the degradation of every important pursuit with which they mix themselves – 'intrigues' in politics and 'friponnes' in science" (Richards 2017, 256). Yet note the sex of the author of the poem just quoted, not to mention Emily Dickinson. George Eliot and her consort, George Lewes, were a little grumpy when the *Origin* appeared, rather putting it down: "We began Darwin's work on 'The Origin of Species' tonight. It seems to be not well written: though full of interesting matter, it is not impressive, from want of luminous and orderly presentation" (Harris and Johnson 1998, 82). This did not stop Eliot from using Darwin-influenced notions of chance to structure her early novel *Silas Marner* (Sonstroem 1998). Darwin liked this novel, but apparently less from the debt to him and more from the fact that it had a pretty girl and a happy ending. (Darwin attended some of the Sunday afternoon soirées that George Eliot cohosted with Lewes. The upper-middle-class Darwins, like their God, had a tolerant attitude to those who broke the Seventh Commandment. The very excited Emma Darwin got to meet the great author. The Puritan lower-middle-class Huxley felt able to attend himself, but would not let his wife visit.)

Whatever Eliot may have thought consciously about Darwin's ideas, Henry James (1873) said, somewhat sourly: "*Middlemarch* is too often an echo of Messrs. Darwin and Huxley" (428). And there were more women writing in a Darwinian mode – Edith Wharton in fiction, for a start. This in itself is hardly any great shock. Nor is the quality. Few if any equal George Eliot in fiction and Emily Dickinson in poetry. Hence, one should be careful about downgrading or belittling their creative productions compared to "real science." Who would at once rank Thomas Henry Huxley above George Eliot, even though to a great extent one is comparing chalk to cheese? Better to think of different aims and tasks. Comparative evaluations are not very helpful. A great science museum may not be an exercise in professional science, but it is of great value in its own right. Before you disagree, visit first the Natural History Museum in London and the Field Museum in Chicago. Visit the Creationist Museum just south of Cincinnati. You may not agree with its message – I don't! – but you must admire the skill with which it is constructed. I don't much care for the coronavirus, but, from a Darwinian perspective, it is a great success.

Through the Twentieth Century

As we move through the twentieth century, one thing seems so obvious as hardly needing us to draw attention to it. Whereas for fifty years after the *Origin* evolution was going to be something of very great importance and interest to people, for the next 100 years, to the present day indeed, other things were going to interest people, including creative people such as novelists and poets. Philosophical ideologies, Marxism, and fascism to give two examples. George Orwell's *1984* is clearly grappling with the latter. T. S. Eliot's poem "The Hollow Men" (1925) is more metaphorical, but it is certainly influenced by the futility and sheer evil of the Great War. All for nothing:

> This is the way the world ends
> This is the way the world ends
> This is the way the world ends
> Not with a bang but a whimper.

Does evolution/Darwinism get squeezed right out? By no means. D. H. Lawrence was a Bergson enthusiast, finding the *élan vital* at work in the blood. In *The Rainbow* (1915) he writes: "He transferred to her the hot, fecund darkness that possessed his own blood" (446); "She had the potent dark stream of her own blood" (449); "The rainbow was arched in their blood and would quiver to life in their spirit" (496). And so on and so forth. The dark side points, as one might anticipate, to rather rude sexual practices – giving strong evidence of why Lawrence is so beloved by adolescents and then goes unread when one reaches adulthood:

> But still the thing terrified him. Awful and threatening it was, dangerous to a degree, even whilst he gave himself to it. It was pure darkness, also. All the shameful things of the body revealed themselves to him now with a sort of sinister, tropical beauty. All the shameful, natural and unnatural acts of sensual voluptuousness which he and the woman partook of together, created together,

> they had their heavy beauty and their delight. Shame, what was it? It was part of extreme delight. It was that part of delight of which man is usually afraid. Why afraid? The secret, shameful things are most terribly beautiful. (237)

In true Bergsonian fashion, this gives rise to ultimate creativity. Writing of the hero of the novel: "And gradually, Brangwen began to find himself free to attend to the outside life as well. His intimate life was so violently active, that it set another man in him free. And this new man turned with interest to public life, to see what part he could take in it" (238). It is with new understanding that I think of the active people I have met in this life.

Bergson continued to fascinate novelists. From the start of his early conversion to Catholicism, the English novelist Graham Greene (1904–91) labored, through the often tortured world of his fictional characters, to make sense of God, of humans, of their relationships. How do we reconcile the evil that lies in people's hearts, and the terrible things that they do, with the existence of a loving Creator God? In later life, by his own admission, Greene was much attracted to the theology of Teilhard de Chardin (1955), although his thinking also drew on process theology reaching back to Alfred North Whitehead (1929), whose God is here with us now, working alongside us and who has human-like features including suffering with us. Synthesizing, Greene saw evil not so much as a consequence of a world, thanks to Darwin, bereft of a God, but something to be conquered and transcended in an evolutionary fashion by humans working with God. Evolution is not the problem; it is the solution.

This world picture infuses Greene's late novel *The Honorary Consul*. A character – almost a parody of a Greene figure, a priest who has married and now joined a revolutionary group who wanted to seize the Argentinian American ambassador and force the Paraguayan dictatorship to hand over captured rebels (and who even more typically has seized the wrong person) – explains the Teilhard/process-influenced theology that drives him forward: "He made us in His image – so our evil is His evil too. How could I love God if He were not like me? Divided like me. Tempted like me" (Greene [1973] 1974, 260). Continuing:

> The God I believe in must be responsible for all the evil as well as all the saints. He has to be a God made in our image with a night side as well as a day side … It is a long struggle and a long suffering, evolution, and I believe God is suffering the same evolution that we are, but perhaps with more pain. (261)

Then:

> God when he is evil demands evil things. He can create monsters like Hitler. He destroys children and cities. But one day with our help he will be able to tear his evil mask off forever. How often the saints have worn an evil mask

> for a time, even Paul. God is joined to us in a sort of blood transfusion. His blood is in our veins and our tainted blood runs through his. (262–63)

For Greene, the priest's theology is not pessimistic, a tragic ending to the Christian story. To the contrary, it points to the triumph of God. For Teilhard de Chardin (1955) evolution is an upwards process ending with the Omega Point. Jesus Christ, our Savior. Success.

> I believe in the Cross and the Redemption. The Redemption of God as well as of Man. I believe that the day side of God, in one moment of happy creation, produced perfect goodness, as a man might paint one perfect picture. God's intention for once was completely fulfilled so that the night side can never win more than a little victory here and there. (Greene 1973, 260)

Increasingly drawing on process theology, where God and humans labor, suffer, and triumph together, Greene continues:

> With our help. Because the evolution of God depends on our evolution. Every evil act of ours strengthens his night side, and every good one helps his day side. We belong to him and he belongs to us. But now at least we can be sure where evolution will end one day – it will end in a goodness like Christ's. It is a terrible process all the same, and the God I believe in suffers as we suffer while, he struggles against himself – against his evil side. (Greene 1973, 262)

For Greene, evolution helps to support and flesh out his conception of God. For the American poet Philip Appleman (1926–2020), evolution had opposite implications. It is the means whereby we can show the moral degeneration at the heart of Christianity. Richard Dawkins, when asked why he found Roman Catholicism so evil, replied that children were brought up thinking it to be true. Appleman, equally opposed to the church as a church, homed right in on the sexual abuse of young children by Catholic priests.

> The first time?
> So long ago—that brown-eyed boy ...
> How can I say this, Your Reverences,
> you'll understand? Maybe
> it was the tilt of his pretty neck
> when he pondered the mysteries—Grace,
> the Trinity—the way his lower lip
> curled like a petal, the way ...
> But you know what I mean—down
> from your pulpits and into the dirty streets—
> you know, there are some provocations
> the good Lord made no sinew
> strong enough to resist.

Surely the love of David for Jonathan excuses this kind of behavior:

> What? I? "Ruined their lives"?
> Wait a minute, let's get this straight—
> my passion gave them a life, gave them
> something rich and ripe in their green youth,
> something to measure all intimate flesh against,
> forever. After that,
> they ruined their own lives, maybe.
> But with me they were full of a love
> firmer than anything their meager years
> had ever tasted …

Away with hypocrisy. The end to this story is known to all:

> Oh, I know where I'm headed—
> to "therapy" as we always say,
> a little paid vacation
> with others who loved not wisely
> but too young—and also, of course,
> with the usual slew of dehydrating
> whiskey priests. But don't forget
> that when they say I'm "recovered" again,
> they'll send me off to another parish,
> with more of those little lambs—a priest,
> after all, is a priest forever.

(Appleman 1996. Permission to reproduce, courtesy University of Arkansas Press)

Religion is evil and its offerings are corrupt. Appleman was the editor of the *Darwin: Norton Critical Edition*. Since this has itself gone through three editions, one assumes he has had a huge influence on the average American undergraduate's understanding of Darwin and his work. Pertinent here is that Appleman saw Darwin's thinking as a cleansing antidote to the moral corruption of Roman Catholicism. True happiness, true joy

> does not depend upon mysticism or dogma or priestly admonition. It is the joy of human life, here and now, unblemished by the dark shadow of whimsical forces in the sky. Charles Darwin's example, both in his work and in his life, helps us to understand that that is the only "heaven" we will ever know. And it is the only one we need (Appleman 2014, 69).

Into the Twenty-First Century: Pro-Darwin

The battles continue, down to the present. Ian McEwan's novel *Enduring Love* (1997) tells a somewhat melodramatic story of six men trying to handle an out-of-control balloon. It starts to rise, and five of the six let go immediately. The sixth hangs on and eventually falls to his death. Two of the survivors discuss the incident. Joe, a non-believing science writer (and a keen Darwinian), is confronted by Jeb, a Christian intent on converting him. Joe fights back:

> No failure. So can we accept that it was right, every man for himself? Were we all happy afterwards that this was a reasonable course? We never had that comfort, for there was a deeper covenant, ancient and automatic, written in our nature. Co-operation – the basis of our earliest hunting successes, the force behind our evolving capacity for language, the glue of our social cohesion. Our misery in the aftermath was proof that we knew we had failed ourselves. But letting go is in our nature too. Selfishness is also written on our hearts. This is our mammalian conflict – what to give to the others, and what to keep for yourself. Treading that line, keeping the others in check, and being kept in check by them, is what we call morality. (14)

Cooperation and selfishness, both products of our evolutionary biology, fusing to produce morality. Thomas Henry Huxley is not where we seek understanding. For him, morality is a nonbiological force keeping biologically caused selfishness in check. As critics pointed out, as visits (or nonvisits) to afternoon tea with George Eliot confirmed, God or no God, Huxley had all the trappings of the deeply conservative Calvinist. "His whole temper and spirit is essentially dogmatic of the Presbyterian or Independent type, and he might fairly be described as a Roundhead who had lost his faith" (Baynes 1873, 502). Charles Darwin is the right starting point. Joe is explicitly endorsing the position of the *Descent*. Morality is an uneasy compromise between two opposing evolutionary instincts that doesn't always work perfectly. But whoever thought that natural selection produces rational perfection? As long as it does better than different solutions – we might all be totally self-serving Scotsmen on the make – we are stuck with the morality we have.

In a later novel, *Saturday* (McEwan 2006) – dealing with the social disruptions that followed the 9/11 plane attack on the Twin Towers – everything is put into an explicitly Darwinian context, trying to make sense of a life/society that seems so harsh and pointless:

> Kindly, driven, infirm Charles in all his humility, bringing on the earthworms and planetary cycles to assist him with a farewell bow. To soften the message, he also summoned up a creator in later editions, but his heart was never really in it. Those five hundred pages deserved only one conclusion: endless and beautiful forms of life, such as you see in a common hedgerow, including exalted beings like ourselves, arose from physical laws, from war of nature, famine and death. This is the grandeur. And a bracing kind of consolation in the brief privilege of consciousness. (56)

Contrast Shakespeare (*Macbeth*), centuries before:

> Life's but a walking shadow, a poor player
> That struts and frets his hour upon the stage
> And then is heard no more: it is a tale
> Told by an idiot, full of sound and fury,
> Signifying nothing.
>
> (Act V, Scene V)

Macbeth saw life as meaningless, whereas Darwin, working with the same (or worse) material, saw that one can justifiably pull an optimistic vision (even if fleeting) from the detritus.

Into the Twenty-First Century: Anti-Darwin

And now to our final novelist, Marilynne Robinson. What I want to stress is, whether or not you accept the philosophy/theology underlying her books – and in many respects I do not – as these words of Barack Obama (in conversation with Robinson towards the end of his presidency) show full well, Darwin produced work that still engages some of the best minds of our generation.

> Well, as you know – I've told you this – I love your books. Some listeners may not have read your work before, which is good, because hopefully they'll go out and buy your books after this conversation.
>
> I first picked up *Gilead*, one of your most wonderful books, here in Iowa. Because I was campaigning at the time, and there's a lot of downtime when you're driving between towns and when you get home late from campaigning. And you and I, therefore, have an Iowa connection, because *Gilead* is actually set here in Iowa.
>
> And I've told you this – one of my favorite characters in fiction is a pastor in *Gilead*, Iowa, named John Ames, who is gracious and courtly and a little bit confused about how to reconcile his faith with all the various travails that his family goes through. And I was just – I just fell in love with the character, fell in love with the book, and then you and I had a chance to meet when you got a fancy award at the White House. And then we had dinner and our conversations continued ever since. (Obama and Robinson 2015)

Marilynne Robinson – author of the trilogy *Gilead* (2004), *Home* (2008), and *Lila* (2014) – is as unfriendly to Darwin as McEwan is friendly. "Whether Darwin himself intended to debunk religion is not a matter of importance, since he was perceived to have done so by those who embraced his views. His theory, a science, is irrelevant to the question of truth of religion. It is only as an inversion of Christian ethicalism that he truly engages religion" (Robinson [1998] 2005, 36–37). It is anti-religion generally (38). It is itself a religion. "Faith is called faith for a reason. Darwinism is another faith – a loyalty to a vision of the nature of things, despite its inaccessibility to demonstration" (39).

Robinson starts with the fundamental premise for all Christians: "Well, I believe that people are images of God. There's no alternative that is theologically respectable to treating people in terms of that understanding" (Obama and Robinson 2015). Does the author make her case, in the context of being a novelist rather than a professional philosopher or theologian? Focus on the major story she tells. The structure is simple. An aged Protestant minister, John Ames, widowed and lonely for forty years, meets half-his-age Lila (who has a somewhat varied background, including prostitution), they fall in love, marry, and have a child, now about seven years old. Central to this tale is the Calvinist notion of predestination. This is in complete contrast to the blind randomness Thomas Hardy extracted from Darwinism. For Calvin, God's sovereignty is everything, meaning that He knows all things and He decides all things. All is the will and scheme of the Almighty. And if you are in any doubt about any of this, in both *Gilead* and *Home* there are explicit discussions of predestination. It is accepted totally, although – in a passage identical in both novels – Ames admits its full meaning is beyond his ken: "I tell them there are certain attributes our faith assigns to God: omniscience, omnipotence, justice, and grace. We human beings have such a slight acquaintance with power and knowledge, so little conception of justice, and so slight a capacity to grace, that the workings of these great attributes together is a mystery we cannot hope to penetrate" (Robinson [2004] 2005, 171; 2008, 220).

If God is so all-powerful, how then does one make room for the equally important matter of human freedom? This is the major intent of Robinson, offering an alternative to the position that many think has been dictated by Darwinism, where all is lawbound and hence there is no free will (Burkeman 2021). She wants to show precisely that her characters are free and merit judgment. Organicism over mechanism, the latter the philosophy behind the Darwinism of McEwan. For Robinson, everything is put into – depends upon – the Christian context. Ames's first wife dies. God's work. "The wife never meant to leave and take the child with her" (2014, 56). He lives a subsequent life of loneliness, whether as part of his God-given nature or part of his reaction to the deaths, from which he never recovered. His vow is to remain faithful. His choice. In old age, he meets and marries a much younger woman, who brings great happiness to him, especially when they have a child. All is the will and scheme of the Almighty. Yet again, we have human choice: "I was getting along with the damned loneliness well enough. Then I saw you that morning. I saw your face" (2014, 85). She proposes. "You ought to marry me" (81). "'Yes,' he said, 'you're right. I will'" (82). It is hard to know who is the more surprised. She hurries off, hot with "shame and anger." He reflects: "Of all the crazy things she had ever done." But these were their choices, not God's. If things went wrong, they were to blame. Not God. Nevertheless, God is working His purpose out. Only through Ames's travail can true grace be understood and appreciated. "I guess I have had my time of suffering." But, reflecting on his love for his new wife: "at least I have had enough of it by now to understand that this is true grace" (132). Grace? "Therefore, since we have been justified by faith, we have peace with God through our Lord Jesus Christ. Through Him we have also obtained access by faith into this grace in which we stand, and we rejoice in hope of the glory of God" (Romans 5:1–2). God's unbounded love for us despite – because of – our weaknesses.

Does Robinson succeed in her attempt to provide a Christian alternative to what she thinks are the theological implications of Darwinism? In many respects, in the context of her novels, she does. She does make plausible the unwillingness of Ames to plunge straight back into marriage when his first wife dies. One can understand why his commitment to his relationship with her makes him reluctant to start again. The loving memory, and also a feeling that this is what God has dictated. God has willed that he become a widower. As a Calvinist should he override this? Then, many years later, when the woman he does marry appears on the scene, it seems to be both God's will and a challenge for this long-single, much older man. Should he do this? When he dies, mother and child will be left alone, without support. Ames remembers the story of Hagar and Ishmael, kicked out of Abraham's family because Sarah is now pregnant, and how God cared for them. It gives Ames a dimension of freedom to make his choice.

It is worth noting that, in this Christian context, the tension between everything being lawbound and yet free is not a problem for Calvinists alone. We have seen Bergson – a Jew close to Catholicism – also wrestling with the problem – arguing that invoking the *élan vital* "implies at least a rudiment of choice" (Bergson 1911, 102). Calvinism, however, thanks to the stress on predestination, does highlight the problem. Although, being a bit of a spoilsport, while I agree with Obama about the worth of the novels, I disagree about Calvinism being in contradiction to Darwinism. There is a well-known philosophical position on free will – "compatibilism" – worked out by people like Jonathan Edwards and David Hume, arguing it is only in the context of a law-bound world that the notion of free will makes sense. In other words, a major reason why I am with Obama on the worth of the trilogy is that I believe Robinson shows brilliantly how someone can be lawbound and yet free. It is just that, unlike Robinson, I do not at all think that this counters Darwinism. It is proof that Darwinism works! Someone who does something without cause is crazy, not free. One might respond that although this is true, the real issue is not lawbound versus nonlawbound but rival visions of the meaning of life: Thomas Hardy, thinking post-Darwinian life without meaning, versus (let us say) Aubrey Moore, who found post-Darwinian life incredibly meaningful. "Darwinism appeared, and, under the guise of a foe, did the work of a friend." Real issue or not, Moore is not atypical. Many Darwinians think that, even if there be no absolute progress, life still makes sense. I am one (Ruse 2021b). Mechanists can find Jesus.

But, to repeat what I said earlier, my aim in discussing the thought and work of Marilynne Robinson is not to show her true or false. Rather to confirm that creative thinkers of the highest order are still wrestling with the ideas of an Englishman who published 150 years ago. Robinson's engagement with Darwin shows the legacy of the reaching out and profound influence of *On the Origin of Species*. As science, and as so much more.

10

Social Issues

Methodological Prolegomenon

In the Introduction to this book, I wrote of myself as a "historian of ideas," rather than a "historian of science." By this, as I explained, I meant that I turn to the past to understand the present. In other words, working within the realm of ideas rather than organisms, I use the past in the way of a biological evolutionist to understand the present. Forestalling criticism, I am not committing the mistake of simply assuming that what holds in the realm of biology transfers readily to what holds in the realm of ideas. I judge the worth of the history of ideas in the light of what it does, rather than by the analogy of other areas of inquiry. I am also not committing the sin of "Whiggishness," simply using the past to justify our beliefs and habits of today, because the present is almost by definition better than the past. I do not deny that often – perhaps usually – the present is better than the past. Anyone who grew up in England in the middle of the last century, as did I, knows full well the troubles of trying to make a phone call when away from home. The too-often-fruitless search for a red phone booth. Today, you just access the already installed number on your mobile phone and get on with telling your spouse that the reason why you are running late is not because you stole extra minutes with your lover, but because a student desperate for help appeared at the door to your office and you took off your outer coat and told the student to sit down and pour out their troubles.

Turning now to what I actually do, rather than what I claim to do, in this final full chapter, I shall look briefly at the four different areas listed in my Introduction, where I argue that Darwin's theory helps us to move forward. My four areas are not chosen randomly. I base my discussion on my home state of Florida. Far from atypical, Florida has the virtue (if such it be) of being more explicit than most in its prejudices. First, *foreigners*, especially immigrants. Feelings range from the uncomfortable to the outrightly hostile. Entirely in line, in September 2022 the Governor of Florida tricked a group of immigrants into taking a flight from Florida that deposited them on Martha's Vineyard, the island off the coast of Massachusetts. "We don't want them; you can have them." Second, *race*. Educational institutions, including public universities, are severely restricted in what they can teach about race. In Florida, what they can say (to middle schoolers) is that black people received a "personal benefit" from slavery because they "developed skills" (Mervosh 2023). Third, *sexual orientation*. Schools are not allowed to teach about the existence, let alone the nature, of gay dispositions, and gay teachers are under huge pressures to keep their lives secret, or preferably to resign. The Governor of Florida has an ongoing dispute with the Disney Corporation, the biggest employer and taxpayer in the state, over its open support of its gay employees and its decision to cease giving donations to anti-gay political parties. Fourth, *women*. Florida has enacted laws that forbid abortion after the sixth week of pregnancy. Needless to say, this includes the use of drugs. In the case of women, freedom of choice – for instance about openly carrying loaded guns – does not extend to freedom of choice over their own bodies. As promised in my Introduction, particularly in the light of recent displays of antisemitism, in the United States particularly but also elsewhere – a function of the Israel–Palestine conflict in Gaza that began with the Hamas raid on Israel of October 7, 2023 – given that the discussions of foreigners and of racial differences focus on questions of ingroup versus outgroup, precisely what is at stake in issues involving prejudice against Jews, I shall take the opportunity of adding a codicil addressing this upsetting social issue.

In this chapter, I shall take the topics in turn. In each case, I shall offer a three-part analysis. First, as a guide to the culture within which Darwin was raised and lived, I shall see what the Bible has to say on these topics. We have seen that not everyone by any means was a literalist, but in a country such as Britain with the established Anglican church – established *Protestant* church – the Bible is a good place to begin our discussion. Since this would have been the most widely read translation, I shall quote from the King James Version. Second, I shall see what Darwin himself has to say on the topics – warts and all, as one might say. Third, notwithstanding the attitudes listed in the last paragraph, I shall ask where we stand today and the relevance of our knowledge to improving the well-being of people in society. My central question is if and how the theory of evolution through natural selection is pertinent and helpful in dealing with the issues. One secondary question I shall slip in, repeatedly, is whether Darwin himself had any inkling of the true revolutionary nature of his theory. And whether he approved!

Foreigners

The Bible

The Bible sends a clear message. Some people are ingroup. Others are not. When it comes to this sort of thing, it can equal Nigel Farage, the leader of the Brexit movement:

> When the Lord your God brings you into the land you are entering to possess and drives out before you many nations – the Hittites, Girgashites, Amorites, Canaanites, Perizzites, Hivites and Jebusites, seven nations larger and stronger than you – and when the Lord your God has delivered them over to you and you have defeated them, then you must destroy them totally. Make no treaty with them, and show them no mercy. (Deuteronomy 7:1)

Fundamental is the pact that God makes with Abraham, renewed with his son Isaac, and then with his son Jacob. You acknowledge me as your God, and you will be the "chosen people" (Figure 10.1).

Figure 10.1 God's covenant with Abraham.
God shows Abraham the stars to illustrate his innumerable descendants, by German painter Julius Schnorr von Carolsfeld (1794–1872), engraving, from "Bibel in Bildern" (1851–60).

ABRAHAM
Genesis
12 Now the LORD had said unto Abram, Get thee out of thy country, and from thy kindred, and from thy father's house, unto a land that I will shew thee:

[2] And I will make of thee a great nation, and I will bless thee, and make thy name great; and thou shalt be a blessing:

[3] And I will bless them that bless thee, and curse him that curseth thee: and in thee shall all families of the earth be blessed.

ISAAC
26 [2] And the LORD appeared unto him, and said, Go not down into Egypt; dwell in the land which I shall tell thee of:

[3] Sojourn in this land, and I will be with thee, and will bless thee; for unto thee, and unto thy seed, I will give all these countries, and I will perform the oath which I sware unto Abraham thy father;

JACOB
28 [13] And, behold, the LORD stood above it, and said, I am the LORD God of Abraham thy father, and the God of Isaac: the land whereon thou liest, to thee will I give it, and to thy seed;

To be fair, not everyone was so harsh on foreigners as the writer in Deuteronomy. "There is neither Jew nor Greek, there is neither bond nor free, there is neither male nor female: for ye are all one in Christ Jesus" (Galatians 3:28). And there is the most touching book in the whole Bible, the story of Ruth. A Moabite woman, married to an Israelite and then widowed, refuses to leave her mother-in-law Naomi when the latter returns to her homeland. "Intreat me not to leave thee, or to return from following after thee: for whither thou goest, I will go; and where thou lodgest, I will lodge: thy people shall be my people, and thy God my God" (Ruth 1:16). She marries the rich landowner Boaz and is the great-grandmother of King David. One can be integrated into the chosen people; but, overall, the Bible, especially the Old Testament, is not exactly foreigner friendly.

Darwin

Darwin exhibited the usual Victorian condescension towards foreigners. Entirely typical is a comment he made in 1875, in a letter to Huxley, about a man who supposedly had professed indifference to the animal suffering caused by vivisection. "I am astounded & disgusted at what you say about

Klein. I am very glad he is a foreigner; but it is most painful as I liked the man" (November 1). Expectedly, the Irish did not escape censure: "What devils the low Irish have proved themselves in New York. If you conquer the South you will have an Ireland fastened to your tail. – " (Letter to Asa Gray, August 4, 1863). Darwin refers to the prominent role played by Irish immigrants in the riots in New York City in July 1863 against the drafting of men into the Union army; the riots resulted in the deaths of at least 105 people (McPherson 1988, 609–10). And, then, remember the comment about how the "more civilised" Caucasians "have beaten the Turkish hollow in the struggle for existence" (Letter to William Graham, July 3, 1881).

This said, it is important to note that Darwin is not quite the hidebound prejudiced Victorian that the last paragraph implies. As we shall see when we come to his thoughts on race, a topic which overlaps with his thinking about foreigners, he does open the way to change, stressing that a lot of the differences between us and other folk are cultural rather than biological.

Today

Ingroup versus outgroup. Move to the thinking today. Because of our newly discovered abilities to find and analyze the DNA of long-gone hominins (human ancestors), we know that for about five million years after leaving the African jungle, where our evolution had taken us, proto-humans were hunter gatherers in bands of about fifty. Although there is a lot of speculation about how and why things happen, note the thinking is always in the context of Darwinian evolutionary theory. Why did the apes leave the jungles and become bipedal, for instance? Today's thinkers are not ancient Greeks, thinking in terms of our bodies reaching up to the divine. Rather, they think in terms of drought drying up the jungles and making for more prairie-like land, and then the virtues of being bipedal – able to look further to spot prey and predators, not to mention the ability to cover long distances. Perhaps physiology is involved: "Overheating is an ever-present problem in the tropics, and by standing up you present less of your body area to absorb the sun's rays" (Tattersall 2022, 82).

Our ancestors began leaving Africa for Europe and Asia long ago: "the earliest well-documented hominin assemblage outside Africa comes from the 1.8 mry-old [million-year-old] site of Dmanisi in the Caucasus" (105). Such invasions continued. These already intelligent primates would always have been on the lookout for new opportunities and sources of food and shelter and the like (Reich 2018). Look now at the most recent significant move out of Africa, some 50,000 years ago or a bit earlier. One group went east toward what we now call Asia and another west to what we now call Europe (Figure 10.2). Always on the

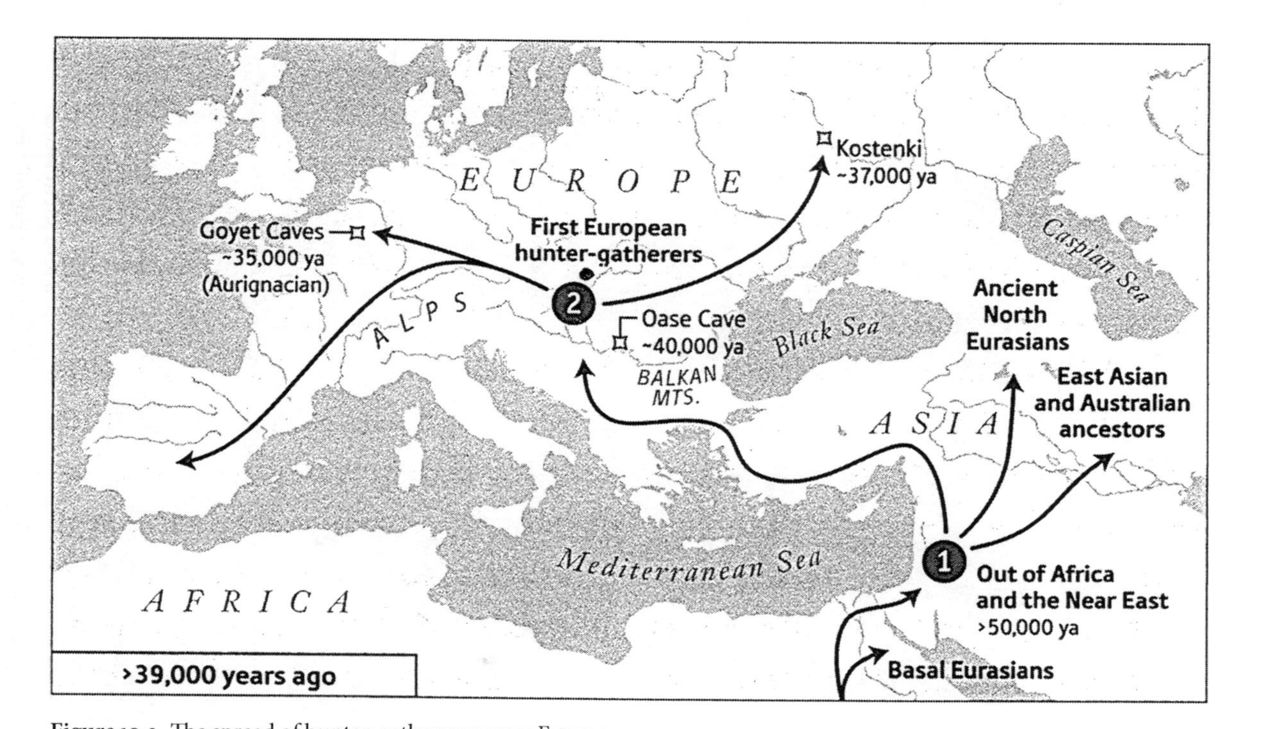

Figure 10.2 The spread of hunter-gatherers across Europe.
(“Map” by Oliver Uberti, copyright © 2018 by David Reich and Eugenie Reich, from *Who We Are and How We Got Here: Ancient DNA and the New Science of the Human Past* by David Reich. Used by permission of Pantheon Books, an imprint of the Knopf Doubleday Publishing Group, a division of Penguin Random House LLC. All rights reserved. UK & Commonwealth (excl. Canada). Brockman, Inc., 260 Fifth Ave., New York, NY 10001)

move, driven in a Darwinian sense to look for better opportunities. Humans are social. A major part of our evolutionary success is that we work together. We achieve our ends by using our brains. Fighting each other means people get hurt, especially organisms which have given up many of their natural adaptations for violence – who would have bet on Muhammad Ali against your local gorilla? – so a major motive behind our nomadic behavior could involve getting away from others in their present surroundings. (I speak with some experience. Facing too many people after too few jobs when I was twenty-two, seeking better opportunities, I moved from England to Canada.)

No sooner had people moved west than an Ice Age set in, making much of northern Europe uninhabitable. People were squashed down into places such as Spain. While you may have little natural inclination to go to war with your competitors, there would be increasing pressure to keep your distance and not let others grab or move in on what you now had – and conversely. As Rudyard Kipling knew full well, wariness about outsiders would be of selective value:

> The Stranger within my gate,
> He may be true or kind,
> But he does not talk my talk –
> I cannot feel his mind.
> I see the face and the eyes and the mouth,
> But not the soul behind.
>
> (Kipling 2001)

This wariness, morphing into hostility, is something that could and would persist as the ice receded and people could start to move north. The invention of agriculture – believed to have started about 10,000 years ago – was something that created assets that others would covet. It was also something which led to huge population increases, and so people started to come into more and more contact with each other, increasing tensions and distrust of strangers (Fry 2013).

Thanks to more efficient methods of agriculture, around 5,000 years ago, there was a major invasion from the east into Europe. The then inhabitants were displaced. Eventually this pushed across to the extremes of the continent, specifically the British Isles. Named after their style of pottery, the "Bell Beaker culture" arrived rather more than 4,000 years ago, and the genetic evidence is that the newcomers threw out the established denizens of the isles (Figure 10.3). With ongoing consequences in our heredity:

> The genetic impact of the spread of peoples from the continent into the British Isles in this period was permanent. British and Irish skeletons from the Bronze Age that followed the Beaker period had at most about 10 percent

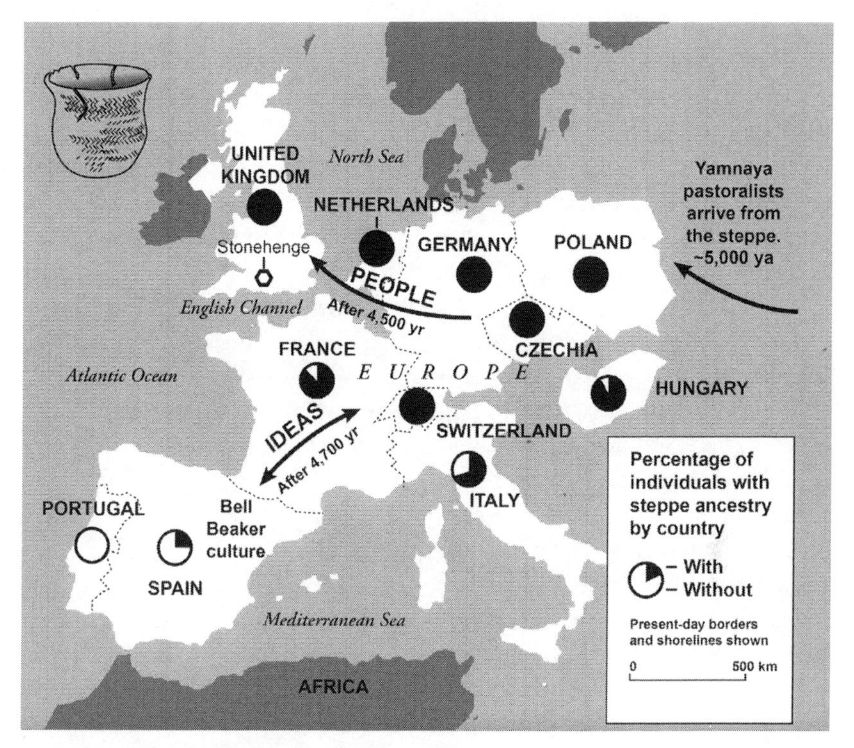

Figure 10.3 The Beaker folk arrive in Britain.
("Map" by Oliver Uberti, copyright © 2018 by David Reich and Eugenie Reich; from *Who We Are and How We Got Here: Ancient DNA and the New Science of the Human Past* by David Reich. Used by permission of Pantheon Books, an imprint of the Knopf Doubleday Publishing Group, a division of Penguin Random House LLC. All rights reserved. UK & Commonwealth (excl. Canada) Brockman, Inc., 260 Fifth Ave., New York, NY 10001)

> ancestry from the first farmers of these islands, with the other 90 percent from people like those closely associated with the Bell Beaker culture in the Netherlands. (Reich 2018, 115)

Since the building of Stonehenge predates this invasion, it means that those who started it and those neo-pagans who today celebrate the Summer Solstice around it have very different ancestries. One can readily see why the different groups involved in this ongoing change would have little love for each other. Conflicts today are simply repeating patterns. In come all these strangers, intent on pushing us to one side. They are not our friends. Think of the several centuries of Europeans in Africa. Or those Europeans who starred in an early chapter pushing out across central America. On reflection, perhaps "starred" is an ill-chosen term. Closer to home, more recent in time, think of the Russians invading East Germany. And, not entirely facetiously, those Polish students taking over behind the counters in Starbucks.

There are three conclusions to be drawn from this. First, we are none of us very different from other humans. In this respect, claims of biological superiority are simply not well taken. There are indeed genetic differences, but as we shall see in the next section, they are not that significant. Second, we may now be in the position where natural selection has eased up. We have seen that Darwin's theory does not insist that everything has adaptive value. Things from the past, such as hostility towards strangers, can persist, so long as there is no strong selective reason why they would be changed and no longer exist. Although now, it may well be in our interests to intervene and change. Else, natural selection could come sweeping back with a vengeance. It may be agreed that there were – and still are – reasons why people are hostile to outsiders. But these are cultural and can be dealt with culturally. There's no need for major enterprises of genetic reengineering. This is not to say it is or will be easy to move forward. Brexit is a prime example of stepping backwards. Stupidity squared and then cubed. Job opportunities notwithstanding, Polish baristas are not about to move into your house and take your husbands and brothers. For all that some of the rhetoric suggests that they are, there are nevertheless real possibilities of change. Third, Charles Darwin would be taken aback. We are about to see evidence that he would not be devastated by thinking that all he had done is totally worthless.

Race

The Bible

Noah gets blind drunk and his son Ham laughs at him:

> 24 And Noah awoke from his wine, and knew what his younger son had done unto him.
>
> 25 And he said, Cursed be Canaan; a servant of servants shall he be unto his brethren. [Canaan was the son of Ham; the curse is on the descendents of Ham.]
>
> 26 And he said, Blessed be the LORD God of Shem; and Canaan shall be his servant.
>
> 27 God shall enlarge Japheth, and he shall dwell in the tents of Shem; and Canaan shall be his servant.

In his *City of God* (2000), the all-influential St. Augustine had some thoughts on this:

> He [God] did not intend that His rational creature, who was made in His image, should have dominion over anything but the irrational creation – not man over man, but man over the beasts. And hence the righteous men in primitive times were made shepherds of cattle rather than kings of men, God intending thus to teach us what the relative position of the creatures is, and what the desert of sin; for it is with justice, we believe, that the condition of slavery is the result of sin. And this is why we do not find the word "slave" in any part of Scripture until righteous Noah branded the sin of his son with this name. It is a name, therefore, introduced by sin and not by nature. This is prescribed by the order of nature: it is thus that God has created man. (19, 15)

Slavery: "It is a name, therefore, introduced by sin and not by nature." Over the years, particularly as Europeans became more aware of Africans, black people were identified as slaves – slaves thanks to sin – a convenient justification of the status of Africans shipped (against their will) to the Americas.

Darwin

The subtitle of Darwin's greatest book is "*the Preservation of Favoured Races in the Struggle for Life.*" One doubts that this means very much. Earlier in the year of publication (1859), he was still mooting using the term "varieties" rather than "races" (Letter to Lyell, March 28, 1859). What does mean very much is that, in common with his family, and indeed with many of his social class, Darwin was a lifelong opponent of slavery. Famous is Darwin's already mentioned maternal grandfather's medallion of a kneeling black slave in chains: "Am I not a man and a brother?" (Figure 2.1). This said, when it came to race, at times Darwin was very much a child of his time. "With civilised nations, the reduced size of the jaws from lessened use, the habitual play of different muscles serving to express different emotions, and the increased size of the brain from greater intellectual activity, have together produced a considerable effect on their general appearance in comparison with savages" (Darwin 1871, 1, 131–32). Moral as well as physical: "Most savages are utterly indifferent to the sufferings of strangers, or even delight in witnessing them. It is well known that the women and children of the North-American Indians aided in torturing their enemies. Some savages take a horrid pleasure in cruelty to animals, and humanity with them is an unknown virtue" (1, 94).

In all fairness to Darwin, one should acknowledge another side to his thinking. He can sound remarkably liberal and modern on the topic of race. For instance, referring to his time on HMS *Beagle*, which carried three Tierra del Fuego natives who had on a previous trip been brought to England, he wrote: "The Fuegians rank amongst the lowest barbarians; but I was continually struck with surprise how closely the three natives on board H.M.S. 'Beagle,' who had lived some years in England and could talk a little English, resembled us in disposition and in most of our mental faculties" (1, 34). In the same mode, Darwin did not seem to be referring to anatomical superiorities when grading different races, but rather, he underlined behaviorally plastic features (Desmond and Moore 2009, 96). Indeed, he emphasized that Western observers were often deceived by slight physical differences between themselves and the "lower" races, and often overvalued those differences:

> Even the most distinct races of man, with the exception of certain negro tribes, are much more like each other in form than would at first be supposed. This is well shewn by the French photographs in the Collection Anthropologique du Muséum of the men belonging to various races, the greater number of which, as many persons to whom I have shown them have remarked, might pass for Europeans. (Darwin 1871, 1, 215–16)

In line with this, Darwin was strongly against polygenism, the suggestion that the races of humans are proof of different species: "Although the existing races of man differ in many respects, as in colour, hair, shape of skull, proportions of the body, &c., yet if their whole organisation be taken into consideration they are found to resemble each other closely in a multitude of points" (1, 231).

Today

Again, ingroup versus outgroup. Take up the topic that lies behind many – most – discussions on race. Where is the evidence that some races/groups are more – or less – intelligent than others? A difficult question to answer because it makes the assumption that it makes sense to talk in the human case of "races." However, if one is to speak of a race or subspecies, you need much more genetic differentiation than one finds between local (human) populations. Generally, if one is to speak of different "races," one looks for at least 25 percent genetic variation between groups (Smith et al. 1997). Hugely striking is that the evidence "confirms the reality of race in chimpanzees using the threshold definition, as 30.1% of the genetic variation is found in the among-race component ... In contrast to chimpanzees, the five major 'races' of humans account for only 4.3% of human genetic variation – well below the 25% threshold. The genetic variation in our species is overwhelmingly variation among individuals (93.2%)" (Templeton 2013). *Homo sapiens* went through bottlenecks. Hence, there simply is not that much genetic variation in our species. Richard Lewontin, as you might expect, seized on this issue:

"The amount of variation that exists within groups vastly exceeds that which exists between them, in whatever characteristic you might care to think of" (Desalle and Tattersall 2022, 133). Although he obviously had ideological reasons for welcoming this conclusion, for once the facts trump ideology.

Not much variation, but there is some variation. As a Darwinian would expect, where it exists systematically, one can look profitably for selection-producing adaptation. For instance, the already mentioned issue of body size/shape: "Within a particular mammal group body form will tend to be linear in hot climates and more compact in cold ones" (132). No big issue as to why this should be so:

> Humans lose heat by the evaporation of sweat from all over their bodies and a linear build is ideal for this, maximizing the heat-losing area of the body relative to its heat-producing interior. In contrast, if you live in a very cold climate you will want to conserve body heat as much as possible, and the ideal form for that would be a sphere, with the smallest area-to-volume ratio. (DeSalle and Tattersall 2022)

The Maasai people are tall, with slender trunks and limbs. The Inuit are much shorter and more stockily built. As is well known, the Maasai live in the hottest part of Africa. The Inuit live in the Arctic. Darwinism vindicated!

And yet where there is selection-fueled variation, intelligence differences are notable for their absence.

> Yes, some humans are "smarter" than others, or better at doing certain things. That is only to be expected. But although "intelligence" is a prized quality that has repeatedly been mentioned in connection with "race," it is something that has proven impossible to measure satisfactorily, not only because it is so complex and multidimensional, but also because it is so tied up with culture, social stratification, and economics. Notions of race do nothing to help clarify this complex situation. (DeSalle and Tattersall 2022, 133–34)

Behind all these discussions about intelligence lie the supposed connection between brain power and skin color. Skin color does not occur by chance. It is adaptive, a function of the distribution of pigment melanin, and from a Darwinian perspective makes perfectly good sense. A darker skin protects from ultraviolet radiation, a big problem in Africa. Especially given that humans have, compared with apes, evolved towards hairlessness, perhaps in part because of Darwin's suggestion of sexual selection, but mainly because sweating (as noted, an important adaptation for animals living out on the savannah rather than in trees) becomes far more efficient: "an explanation based on natural selection for enhanced thermoregulation during high physical activity levels under conditions of high environmental heat load" (Jablonski and Chaplin

2017, 2). However: "Loss of body hair was accompanied by disadvantages, notably, loss of some protection against abrasion and ultraviolet radiation (UVR). Compensatory changes evolved quickly in hominin skin." Humans changed in the direction of dark skins (1.2 mya). Subject to the possibility of reverse change. Remarkably, it seems that in the case of Europeans, the major changes came 12,000 years ago. In the absence of strong sunlight, white skin does a better job of Vitamin D synthesis – invaluable for those ongoing dark days of Northern Europe. Whatever later cultural overlays there may be, we are not talking here about brute intelligence or anything like that. (Asians have light skin, but the genetic mechanisms are different from those of Europeans.)

Jews

We may be prejudiced, but our biology shows that such attitudes based on race – or, more carefully, "race," implying nonexistence – are without foundation. Moreover, although Darwin was caught in the prejudices of the Victorian era, as we also see repeatedly, his theory had the seeds that led to their refutation. Most pertinently – natural selection! And Darwin himself had some suspicions that he himself did not know the whole truth and that today's beliefs might be plausible. Which point is a good opening to add a short codicil about the Jews, for prejudice towards them – antisemitism – is properly judged to be a matter of ingroup versus nongroup, whether this is a matter of nationality or of race or some combination of the two. What one can say is that it is a matter of longstanding and evil intent. In Clifford's Tower, in the City of York, on March 16, 1190, 150 Jews were trapped and massacred. (Most committed suicide rather than be killed.) Coming rapidly down to the present, we have the final solution and Auschwitz. Today, antisemitism has not vanished, as the United States shows depressingly repeatedly, most recently in demonstrations against Israel over its fraught relationships with Arabs.

If biblical justification is sought, the Gospel of St. John – in the context of the Jews urging the Romans to crucify Jesus – is a common resource.

> 8 44 Ye are of your father the devil, and the lusts of your father ye will do. He was a murderer from the beginning, and abode not in the truth, because there is no truth in him. When he speaketh a lie, he speaketh of his own: for he is a liar, and the father of it.
>
> 45 And because I tell you the truth, ye believe me not.
>
> 46 Which of you convinceth me of sin? And if I say the truth, why do ye not believe me?
>
> 47 He that is of God heareth God's words: ye therefore hear them not, because ye are not of God.

Overall, Darwin was not terribly interested in the Jews. We have seen he discussed the implications of circumcision, in common with his class we get the odd comment about Jews and money, and – most notably – he added his signature to a letter protesting the persecution of Jews in Russia:

> To the Right Hon. the Lord Mayor of the City of London.
>
> My Lord:
> We, the undersigned, consider that there should be a public expression of opinion respecting the persecution which the Jews of Russia have recently and for some time past suffered. We therefore ask your lordship to be so good as to call, at your earliest convenience, a public meeting for that purpose at the Mansion House, and that you will be good enough to take the chair on the occasion. (Letter dated January 21, 1882)

Other signatories included the Archbishop of Canterbury and Cardinal Henry Manning.

Darwin was certainly not one who thought that Jews are significantly the "other." In the *Descent* he wrote:

> The singular fact that Europeans and Hindoos, who belong to the same Aryan stock and speak a language fundamentally the same, differ widely in appearance, whilst Europeans differ but little from Jews, who belong to the Semitic stock and speak quite another language, has been accounted for by Broca through the Aryan branches having been largely crossed during their wide diffusion by various indigenous tribes. (Darwin 1871, 1, 240)

Is that still our thinking today? One can certainly agree that Darwin's theory of evolution through natural selection in no way supports antisemitism of the ferocity found in the Third Reich. To the contrary, it furnishes strong evidence against it. To quote what was stated just a little earlier: "First, we are none of us very different from other humans. In this respect, claims of biological superiority are simply not well taken." The same is true of claims of biological inferiority. Granted. But is this enough to negate all possible differences? Shylock, in *The Merchant of Venice*, is recognizably a Jew, and that is not exactly regarded as to his advantage. Indeed, at the end of the play, he converts to Christianity! Unsurprisingly, the play was much performed under the Third Reich.

Does any of this relate to possible Jew–Gentile (biological) differences? As we have just seen, genetic testing does certainly show that there are differences between human groups, and some of them (for instance involving skin color) are as certainly linked to adaptive benefits. Not all cases. The variations, although systematic, might be so minor they really don't have much or less adaptive advantage. A much discussed study found that there are some consistent genetic correlations separating specific groups. For instance, according to lore and language similarity, the somewhat isolated Kalash in northern Pakistan have origins linking them to Europeans, and genetics backs this up. "Genetic clusters often corresponded closely to predefined regional or population groups or to collections of geographically and linguistically similar populations" (Rosenberg et al. 2002, 2384). Not much evidence of significant adaptive advantage. More a matter of slight differences between the founding organisms.

In the case of the Jews, the genetic differences between groups, as well as between Jews and Gentiles, are strong enough that plausible hypotheses can be made about the history of Jewish wandering over the past three millennia. "Early population genetic studies based on blood groups and serum markers provided evidence that most Jewish Diaspora groups originated in the Middle East and that paired Jewish populations were more similar genetically than paired Jewish and non-Jewish populations" (Ostrer and Skoreck 2013, 121). Thus, for instance: "It was observed that the Jewish populations of Europe, North Africa, and the Middle East formed a tight cluster that distinguished them from their non-Jewish neighbors. Within this central cluster, each of these Jewish populations formed its own subcluster, in addition to the more remote localization of members of some Diaspora communities." Expectedly, as hypothesized earlier, the genes often showed that differences were related to the somewhat randomly chosen founders of populations: "Analysis of Jewish mitochondrial genomes in some Diaspora communities has demonstrated limited genetic diversity and therefore, evidence for strong founder effects." Moreover, the timespans of coliving with gentiles gave evidence of rates and sources of interbreeding. "A high degree of European admixture (30–60%) was observed among Ashkenazi, Sephardic, Italian and Syrian Jews. The North African Jewish groups demonstrated North African and Middle Eastern admixture with varying European admixture."

All very interesting (and plausible). Are there any implications of significant differences, possibly adaptive? It is well known that some diseases are more common in some groups than in other Jewish groups, not to mention non-Jewish populations. Jewish physicians and scientists have been the leaders in this work. Tay-Sachs disease, far more common in Ashkenazi Jews, is the best-known example. For the rest, any significant systematic differences are much more plausibly cultural than genetic. There simply has not been enough time for natural section to promote adaptively advantageous features, apart from the fact that Jews and Gentiles have generally similar environments. Disraeli and Gladstone, for instance, both sat on the front benches of the same Houses of Parliament. Culture is perhaps somewhat different and could tear groups apart. To take Darwin's example of Jews and money, if there are differences – Jews more careful with/obsessed by money – better turn to the New Testament for an answer. Usury was forbidden to Christians. Luke 8:

> 34 And if ye lend to them of whom ye hope to receive, what thank have ye? for sinners also lend to sinners, to receive as much again. 35 But love ye your enemies, and do good, and lend, hoping for nothing again; and your reward shall be great, and ye shall be the children of the Highest: for he is kind unto the unthankful and to the evil.

Christians, however, could get Jews to do their dirty business, and so they did.

All told, Darwinian theory gives little support even to relatively mild genetically based Jewish–Gentile behavioral or like differences. It does invite one to explore the role of possible cultural factors.

Sexual Orientation

The Bible

Not much ambiguity here:

> Leviticus 18:22 You shall not lie with a male as with a woman; it is an abomination.
>
> Leviticus 20:13 If a man lies with a male as with a woman, both of them have committed an abomination; they shall surely be put to death; their blood is upon them.
>
> Romans 1:26–28 For this reason God gave them up to dishonorable passions. For their women exchanged natural relations for those that are contrary to nature; and the men likewise gave up natural relations with women and were consumed with passion for one another, men committing shameless acts with men and receiving in themselves the due penalty for their error. And since they did not see fit to acknowledge God, God gave them up to a debased mind to do what ought not to be done.

Darwin

Until recently, Darwin's thinking on sexual orientation has not attracted much attention. It is not particularly hidden, but has never been put together in a coherent whole. Over thirty years ago, I wrote a full-length book on the topic of same-sex feelings and activity, without any discussion of Darwin (Ruse 1988a). Here, I follow closely the path-breaking study by Ross Brooks (2021), to whom all credit is due. From the first, in a private notebook, Darwin noted that human beings still show the traces of both sexes: "Every man and woman is hermaphrodite" (Darwin 1987, D 162; September 16, 1838). The years went by. Darwin kept note of the pertinent research. By the time of the *Descent*, Darwin accepted that homosexual behavior was something that would occur in the human species. Yet candor only goes so far. He certainly didn't want to imply that such behavior, presumably natural in some sense, was going on among civilized folk today and that it was acceptable. Darwin came up with a good Victorian solution. It's all the fault of the savages! "The greatest intemperance with savages is no reproach. Their utter licentiousness, not to mention unnatural crimes, is something astounding" (Darwin 1871, 1, 96). Adding:

> The hatred of indecency, which appears to us so natural as to be thought innate, and which is so valuable an aid to chastity, is a modern virtue, appertaining exclusively, as Sir G. Staunton [an employee of the East India Company] remarks, to civilised life. This is shewn by the ancient religious rites of various nations, by the drawings on the walls of Pompeii, and by the practices of many savages.

Today

Darwin had inserted the wedge against homosexuality as abnormal. The thin end of a very large wedge. No one was fooled. One of Darwin's correspondents pointed out that it is more than savages who indulge in homosexual activities. It was an openly accepted practice in Ancient Greece:

> I know no more instructive fact – disagreeable as it is, it is of high scientific interest – than that one practice (to denote it by the general term I have been using), paiderastia, in many countries became systematised. Thus in Greece the relation between a man and his youthful lover was constituted by a form of marriage after contract between the relatives on both sides. (Letter from John McLennan, February 3, 1874)

One suspects that, thanks to his schoolboy immersion in the classics, none of this was a great surprise to Darwin. I too read the *Symposium*.

Almost to be anticipated, Catholic zoologist and Darwin critic St. George Mivart used the topic to write an excoriating review of a short piece on human sexuality penned by Darwin's son George. In the widely read *Quarterly Review*, Mivart wrote:

> There is no hideous sexual criminality of Pagan days that might not be defended on the principles advocated by the school to which this writer

> [George Darwin] belongs. This repulsive phenomenon affords a fresh demonstration of what France of the Regency and Pagan Rome long ago demonstrated; namely, how easily the most profound moral corruption can co-exist with the most varied appliances of a complex civilization. (Mivart 1874, 70)

The damage was done. The Darwinians went after Mivart with a ferocity that almost suggests that, after all, there is something to the cultural evolution analogy about the struggle for existence. It could not be denied that Darwin's theory suggests that homosexual behavior is universal and hence, plausibly, natural. Yet not everyone found this as morally offensive as Mivart. Jumping forward, at the beginning of the twentieth century, Darwin enthusiast, the English naturalist Edmund Selous, almost reveled in same-sex activity in birds:

> If we say it is vitiated or perverted instinct, still there must be a natural cause for what we regard as the perversion. As is well known, hermaphroditism preceded, in the march of life, the separation of the sexes, and all of the higher vertebrate animals, including man, retain in their organisms the traces of this early state. If the structure has been partly retained, it does not seem unlikely that the feelings connected with it have, through a long succession of generations, been retained also, and that, though more or less latent, they are still more or less liable to become occasionally active. This view would not only explain such actions as I have here recorded, but many others scattered throughout the whole animal kingdom, and might even help to guide us in the wide domain of human ethics. (Selous 1901–2, 182)

"Human ethics"! "Selous's remarks here mark a significant return to, and development of, a vision of sexual development that was implicit in Descent but which Darwin himself had failed to elucidate" (Brooks 2021, 332). If homosexuality is natural, then is it to be condemned as immoral? Our primary emotions should be against people who condemn homosexuality, not those who practice it. Albinos are not condemned on moral grounds.

One who picked up on this was Sigmund Freud. He was, as one might expect of one born in 1856, a fully committed evolutionist, and he acknowledged the *Descent of Man* (along with Copernicus in science and Goethe and Shakespeare in literature) as one of the ten most influential books he had read. By the 1890s, Darwin's influence on Freud's generation, particularly his emphasis on sexual selection, "had become so extensive that Freud himself probably never knew just how much he owed to this one intellectual source" (Sulloway 1979, 275). Particularly significant was the claim that we evolved from small groups of hunter-gatherers: "From Darwin I borrowed the hypothesis that human beings originally lived in small hordes, each of which lived under the despotic rule of an older male who appropriated all the females and castigated or disposed of the younger males, including his own sons" (Freud 1960, 99). We are on the way to the Oedipus Complex: boys wanting to kill their fathers so they can have sex with their mothers. Freud might praise Darwin, but is he doing it more to raise his own status rather than because of a genuine debt? Not so:

> Judging from the social habits of man as he now exists, and from most savages being polygamists, the most probable view is that primeval man aboriginally lived in small communities, each with as many wives as he could support and obtain, whom he would have jealously guarded against all other men. Or he may have lived with several wives by himself, like the Gorilla; for all the natives "agree that but one adult male is seen in a band; when the young male grows up, a contest takes place for mastery, and the strongest, by killing and driving out the others, establishes himself as the head of the community." The younger males, being thus expelled and wandering about, would, when at last successful in finding a partner, prevent too close interbreeding within the limits of the same family. (Darwin 1871, 2, 362–63)

As we all know, over the years, the Oedipus Complex has become for many a derisory source of amusement, a good starting point to discount Freud's thinking in general. It does seem a stretch to suggest that homosexuality is a result of a failed attempt to resolve the complex, taking oneself out of contest with same-sex father by ending sexual desire for the other-sex mother. Be this as it may, something led to Freud's "Letter to an American Mother," written in 1935:

> Dear Mrs .... I gather from your letter that your son is a homosexual. I am most impressed by the fact that you do not mention this term yourself in your information about him. May I question you, why you avoid it? Homosexuality is assuredly no advantage, but it is nothing to be ashamed of, no vice, no degradation, it cannot be classified as an illness; we consider it to be a variation of the sexual function produced by a certain arrest of sexual development. Many highly respectable individuals of ancient and modern times have been homosexuals, several of the greatest men among them (Plato, Michelangelo, Leonardo da Vinci, etc.). It is a great injustice to persecute homosexuality as a crime, and cruelty too. If you do not believe me, read the books of Havelock Ellis. By asking me if I can help, you mean, I suppose, if I can abolish homosexuality and make normal heterosexuality take its place. The answer is, in a general way, we cannot promise to achieve it. In a certain number of cases we succeed in developing the blighted germs of heterosexual tendencies which are present in every homosexual, in the

> majority of cases it is no more possible. It is a question of the quality and the age of the individual. The result of treatment cannot be predicted. What analysis can do for your son runs in a different line. If he is unhappy, neurotic, torn by conflicts, inhibited in his social life, analysis may bring him harmony, peace of mind, full efficiency whether he remains a homosexual or gets changed.... Sincerely yours with kind wishes, Freud.

Whatever the connection between Freud's theories and the message of the letter – "arrest of ... development" is an uncomfortable sentiment – it is true that the letter was very much in the mode of a Darwin-influenced thinker. Homosexuality is a matter of biology and not a matter of choice. Of course, having been produced by natural selection does not at once imply that one will be happy. There is certainly a presumption that, if homosexuals are unhappy, it is not because of their sexual orientation, but because society makes them unhappy. It is our fault as much as anyone's.

One obvious question has yet to be answered. If homosexuality is natural – note, until recently, as with Darwin, the focus was on males – and if we are at the least dubious about the role of Oedipus Complexes (assuming that they do exist), what possible Darwinian reason could there be for this? Men having sex together do not produce babies. The possible causal role of natural selection is, as they say, highly contested (Ruse 1988a). Repeated studies estimate the number of male homosexuals at around 3–4 percent; females are somewhat less. This is not to say that being homosexual means that one will not have children, but the Kinsey studies show that orientation (especially of males) is a very significant (behavioral) factor in having fewer offspring (Bell and Weinberg 1978). Presumably, one could link things back to those Darwinian alpha males crushing all opposition. But it is hard to see why the losers would turn homosexual or have the genes to pass this on.

As a first step, it is probably a good strategy to turn first to homosexuals today. When thinking of proximate causes, most researchers today think in terms of hormones, particularly as they affect fetal brain development. Comparative levels of testosterone during the third to sixth months of hypothalamus development seem to be the all-important factors (LeVay 2010). Could the selective (ultimate) cause be kin selection, with gay or lesbian siblings helping other family members to reproduce? Could it be a case of heterozygote fitness, where the heterozygote has more offspring, balancing the fact that the homozygotes have fewer offspring? Could it be "parental manipulation," with the mother's biology kicking in to control the reproduction of her offspring? If resources are limited, it might not be a good thing to have all the sons competing equally. Generally, as one moves down to the bottom of the birth order, the higher the incidence of homosexual orientation. At the proximate level, it could be that the more boys, the more the female levels of prenatal hormones get changed. Then, perhaps, at the ultimate level, selection picks up and makes use of this consequence. Whatever the case, it does seem plausible that homosexuality is "natural" – there is no need to invent up some fancy new mechanism – and there is no reason at all to think it a deviancy, like psychopathy. Darwin may not have been comfortable with this conclusion. Overwhelmingly, his science pointed to this conclusion.

Facts like these have led to change, if slowly. England is the country that sent Oscar Wilde to jail (1895) because of his relationship with Lord Alfred Douglas. Freud's contemporaries did not always agree with him about homosexuality being, in some important sense, normal. Half a century after Wilde, nearly two decades after Freud's letter, Franz Kallmann, noteworthy because of his twin studies, wrote that homosexuality is "an inexhaustible source of unhappiness, discontent and a distorted sense of human values" (1952, 296). At the same time, as readers of the (English) Sunday newspaper *News of the World* knew well, the police were extremely active in enforcing the law against (male) homosexual activity. It was called the "blackmailer's charter," because, when a wretched victim drew attention to his predicament, not only was the blackmailer prosecuted, but also the victim. In 1954 over 1,000 gay men were in jail because of their illegal activities. For all that, as critics pointed out, sending a homosexual to prison was about as effective as sending a drunkard to a brewery. The law was not repealed until 1967. Slowly, homosexuals of both sexes (as well as others under the category LGBT) were granted the rights of heterosexuals. Civil unions were allowed from 2005 and same sex marriage from 2014. (A similar tale can be told of the United States.) In 1995, exactly one hundred years after he was sent to jail for homosexual behavior, a memorial window to Oscar Wilde was unveiled in Poet's Corner in the resting place of Charles Darwin and Isaac Newton, Westminster Abbey.

The pope's recent pronouncements warn us that we still have a way to go: "It is a sin, as is any sexual act outside of marriage" (Winfield 2023). Protestants are little better: "A quarter of U.S. congregations in the United Methodist Church have received permission to leave the denomination during a five-year window, closing this month [December 2023], that authorized departures for congregations over disputes involving the church's LGBTQ-related policies" (Smith 2023). To be fair, many Christians believe "we are all one in Christ Jesus." Change is possible. Change has occurred. We are not yet at journey's end. Darwin deserves full credit for his role in helping us on our way.

Women

The Bible

As with homosexuality, there is not much ambiguity on this topic, starting with the fact that the reason why humans are always in such a mess is that Eve could not resist the apple and seduced Adam into going along with her. Fig leaves all around! An attitude that persisted. St. Paul on women:

> 34 Let your women keep silence in the churches: for it is not permitted unto them to speak; but they are commanded to be under obedience as also saith the law.
>
> 35 And if they will learn any thing, let them ask their husbands at home: for it is a shame for women to speak in the church.

Even today there are biblically based decisions that are breathtaking in their belittlement of the status of women. The Southern Baptists, America's largest Protestant denomination – over fourteen million members, 85 percent white – have voted overwhelmingly to expel those churches that have permitted female pastors (Dias and Graham 2023). This includes the megachurch Saddleback, founded by Rick Warren, author of the best-selling *The Purpose Driven Church*. We seem to have a story that has been made up by a New Atheist. Not so, I am afraid.

Darwin

Darwin shows himself to be very Victorian on this topic. "Man is more courageous, pugnacious, and energetic than woman, and has a more inventive genius. His brain is absolutely larger, but whether relatively to the larger size of his body, in comparison with that of woman, has not, I believe been fully ascertained" (Darwin 1871, 2, 316–17). And much more along the same lines. Women just cannot win. "The chief distinction in the intellectual powers of the two sexes is shewn by man attaining to a higher eminence, in whatever he takes up, than woman can attain – whether requiring deep thought, reason, or imagination, or merely the use of the senses and hands" (2, 327). If one drew up a list of six male and six female philosophers, "the two lists would not bear comparison." (Darwin came a bit early for B. J. B. Lipscomb's *The Women Are Up to Something: How Elizabeth Anscombe, Philippa Foot, Mary Midgley, and Iris Murdoch Revolutionized Ethics*, published in 2021.)

Today

Expectedly, Darwin drew his fair share of scorn. *The Evolution of Woman: An Inquiry into the Dogma of Her Inferiority to Man*, by Eliza Burt Gamble

(1894), excoriated Darwin for suggesting that the larger size of men over women pointed to their superiority, when he himself argued that the larger size and strength of gorillas over humans precludes their being fully social and hence points to their inferiority! However, take note that, whatever Darwin's personal views – and we shall see reason shortly to think they were more sophisticated than the impression just given – as many Victorian-era feminists pointed out, if Darwin be true then Adam and Eve are not, and in one stroke the theological basis for male superiority is gone (Hamlin 2014). "Woman can no longer be taunted with having brought on humanity the traditional curse" (37). Indeed, ignoring Darwin's worries about progress, turning claims that Eve's being made from Adam's rib shows her second-rate status, Darwin's theory was used to support female superiority (Figure 10.4): "If we find God gradually advancing in his work from the inorganic earth to the mineral kingdom, then to the vegetable kingdom, and last of all making man, the fact that woman is made after man suggests her higher qualities rather than man's superiority" (39). A similar sentiment:

> It was not, however, until the year 1886, after a careful reading of *The Descent of Man*, by Mr. Darwin, that I first became impressed with the belief that the theory of evolution, as enunciated by scientists, furnishes much evidence going to show that the female among all the orders of life, man included, represents a higher stage of development than the male. (Gamble 1894, v–vi)

Turn now to the present. What does science tell us today? Go back to what Freud seized on, what has, jocularly, been referred to as our five-million-year camping trip. Small bands of hunter-gatherers, proto-humans. The evidence both from present-day groups and from archaeology is that females played an active role. One hates to disillusion the adolescents of my generation – not to mention Darwin and Freud – but the typical *National Geographic* picture of dominant males, clad only in athletic supports, armed with spears for the hunt, with the little bare-breasted women staying home and looking after babies, is just not true. Humans are not lions or tigers. As noted earlier, we have gone the route of sociality and intelligence, rather than high speed to hunt down our prey and fangs and claws to kill our catch. We rely on artifacts – traps and the like – not to mention the knowledge of how best to use them. Women are equals. Likewise, when it comes to food processing and the like, males and females both have stakes in working efficiently (Ruse 2022b).

Crucially, however dominant males may seem, as with the chimpanzees, as those Victorian feminists kept insisting, men need female support: "Women are not accorded a lower status due to their childbearing, childrearing, and lactating functions, but rather, are honoured by men for these contributions"

Figure 10.4 God creating Eve from Adam's rib. From *Liber chronicarum mundi*, 1493 (Nuremberg Chronicle).

(Jarvenpa and Brumbach 2014, 1253). Ongoing rape is simply not the best way of getting sexual favors and thus producing children, especially sons. Getting along and cooperating is a much better strategy:

> A possible clue for the evolution of sex equality in the hominin lineage was the increase in the cost of human reproduction associated with larger brain sizes in early Homo. Higher offspring costs would require investment from both mothers and fathers, as seen among extant hunter-gatherers. The need for biparental investment predicts increased sex equality, which is reflected in the high frequency of monogamy and the reproductive schedules of male

> hunter-gatherers who typically stop reproducing early and exhibit long life spans after their last reproduction. (Dyble et al. 2015, 798)

Never forget the power that women have over their children. Freud spoke not just of dominant males but also of the importance of mothers for sons. Treating women as dirt is simply not a good evolutionary strategy. Groups must have females, and natural selection is going to promote their value – as producers of more group members and as contributing to the whole (Hrdy 1999).

Female inequality is not biologically engrained. It does not necessarily exist. And yet it does. How do we get the kinds of forces that led to so many still belittling women and their abilities and roles in life? The answer presses itself on us. Inequality is in major part a function of the already mentioned move to agriculture about 10,000 years ago. Among farmers, hunter-gatherer pressures are off and males can more readily manipulate themselves into power. The traditional "pattern contrasts with that of male farmers and pastoralists, whose reproductive spans extend well into late life. The recognition of affinal ties throughout our long life span has been argued to be an important step in human social evolution" (Adovasio, Soffer, and Page 2007, 269). If, thanks to agriculture, women go on having many offspring, then it follows that they are going to be tied down to the basic needs of infants and small children. One does not have to be the worst kind of chauvinist to recognize the sexes are different and relabeling has limited power. Men do not breast feed. Gender differences will appear and be accentuated. Men will have a dimension of freedom that women did not have. "With more pregnancies, women had to spend more energy on nurturing zygotes, fetuses, and helpless babies – a costly enterprise indeed" (Adovasio, Soffer, and Page 2007, 269–70).

The obvious implication is that, if women are freed from culturally imposed tasks, there is absolutely no reason at all to think they will prove less intelligent or able to control and direct things. Plotting the expected path of an antelope demands the ability to think and then to put these thoughts to full use. Designing and making traps for smaller mammals seems to demand no less ability to think and then to put the thoughts into action. Certainly, anyone who has been in universities in the past half-century can and will tell you that, when they are given the chance, young women are as good at if not better than young men. "Consider that by age 25, over one-third of women have completed college (versus 29% of males); women outperform men in nearly all high school and college courses, including mathematics; women now comprise 48% of all college math majors; and women enter graduate and professional schools in numbers equal to most, but not all fields" (Ceci and Williams 2009, 5).

What brought this about? Two obvious factors. First, at least in the West, machines have transformed women's lives. Formerly, the week was dominated by washing by hand in the tub – Monday – mangling and hanging the clothes out to dry whenever there is a trace of sunshine – Tuesday – and then ironing – and more ironing – followed by folding and putting in the airing cupboard – Wednesday, Thursday, Friday, Saturday. No more! Thanks to the Bendix washing machine company – it first started making washers in 1938 – lives filled with drudgery are but a memory. Hours spent over soapsuds could be replaced by hours over philosophical tracts. (Don't laugh. That is exactly the story of my mother. The child is the mother of the man.) Women show Darwin! Another discovery freeing women and making them much more equal and ready to compete with men was the coming of the "pill" in the 1960s. This sparked changes in the succeeding years: The status of young women changed from vulnerable creatures needing protection from savage predators – "even the nicest boys want only one thing" – to equals, socially and sexually. Humans, more than ready to take the forward, dominant role.

Humans are intelligent, social beings. Whatever the personal prejudices of evolutionists, including Darwin himself, despite his proto-Freudian musings about the male-dominated nature of groups, there is nothing in his actual theory that says, for the optimal functioning of a group, males must be dominant over females. No doubt in a postagricultural society, as with war, there will be factors that bring this on. A much increased number of children on whom women are by necessity obliged to focus. But as and when these factors change or are reduced or eliminated – machines and contraceptives – so the inevitability of male dominance will change or be reduced or eliminated. Women can regain their earlier status (Hamlin 2014, 100).

Last Words

One final point before leaving the discussion of the status and role of women. We have seen repeatedly that there are overt seeds in Darwin's own thinking to suggest that, for all he talked like a typical Victorian, he was aware of the revolutionary implications of his selection theory. Despite the incredibly sexist writings on women, the seeds are there too. Adam and Eve are not major elements of Darwin's world picture. Moreover, modifying opinions based on the public Darwin – the male sexist – looking at Darwin's private correspondence leads to a somewhat different take on things. No one is going to claim that Darwin was a secret admirer of Gloria Steinem, but he was not the kind of uncritical and enthusiastic women-belittler too frequent today. Even ignoring the fact that the Southern Baptists came into being because (unlike their northern coreligionists) they endorsed slavery, Darwin would have been out of place given his full views on the status of women. In 1871, the year of the *Descent*, to one correspondent he wrote:

> **Madam**
> I have the honour to acknowledge, on the part of Mrs Darwin & myself, the request that we should agree to our names being added to the General Committee for securing medical education to women.
> I shall be very glad to have my name put down, or that of Mrs Darwin but I should not like both our names to appear.
> With sincere good wishes for the cause you are so generously aiding I beg leave to remain | Madam | your obedient servant | Charles Darwin (Letter to Louisa Stevenson, April 18, 1871)

Stevenson was an honorary secretary of the Committee for Securing a Medical Education to the Women of Edinburgh. Darwin's name duly appeared (Roberts 1993, 97). One suspects Emma Darwin may have had a hand in this matter. (To be fair to Huxley, as time went by, he did teach some physiology to women studying to become doctors. But only in segregated classes and as job training, not pure research.)

Truly interesting is an exchange between Darwin and another correspondent, an American woman, late in his life (Saini 2017). She had just heard a paper that claimed that women were fated always to be man's intellectual inferiors and the *Descent* was used in support of the argument. Was he being cited properly, and if so, did Darwin still believe this?

> As a believer in continued scientific discoveries and revelations answering and modifying, ultimately, all material questions; and as an admirer of your cautious and candid methods of conveying great results of learning and investigations to the world, I take the liberty to inquire whether the Author of the paper rightly inferred her arguments from your work: or if so, whether you are of the same mind now, as to possibilities for women, judging from her organization &c. (Letter from C. A. Kennard, December 26, 1881)

Darwin replied that he was cited correctly, but there is a bit more to the story:

> The question to which you refer is a very difficult one. I have discussed it briefly in my "Descent of Man". I certainly think that women though generally superior to men in moral qualities are inferior intellectually; & there seems to me to be a great difficulty from the laws of inheritance, (if I understand these laws rightly) in their becoming the intellectual equals of man. On the other hand there is some reason to believe that aboriginally (& to the present day in the case of Savages) men & women were equal in this respect, & this wd. greatly favour their recovering this equality. But to do this, as I believe, women must become as regular "bread-winners" as are men; & we may suspect that the early education of our children, not to mention

> the happiness of our homes, would in this case greatly suffer. (Letter to Kennard, January 9, 1882)

At once Kennard replied, that she agreed with Darwin on the higher moral nature of women! "I believe you are supported in your ideas of the greater moral qualities of woman." She did take exception to the claim about "bread-winners" and the ill effects on children of such being the case for women.

> And why be anxious for the "education of our children" and "the happiness of our homes", if women become bread winners? when in this country five sixths of the educators are women and acknowledged "breadwinners", beside improving the condition of their homes and adding happiness thereto –
>
> Which of the partners in a family is the breadwinner where the husband works a certain number of hours in the week and brings home a pittance of his earnings (the rest going for drinks & supply of pipe) to his wife; who, early & late, with no end of self sacrifice in scrimping for her loved ones, toils to make each penny tell for the best economy and besides, to these pennies she may add by labor outside or taken in? (Letter from Kennard, January 28, 1882)

I suspect that anyone who has, like me, been in the education business for half a century, will be inclined to agree with Kennard. I taught at a university with a veterinary college. When I started, in 1965, the incoming class was eighty men and a quota of four women. When I left in 2000, the incoming class was 100 students, over 90 percent of whom were women. (The TV series *All Creatures Great and Small* meant that every little girl in Canada wanted to be a vet. Many did.)

Envoi

Charles Darwin was no rebel. He exhibited every prejudice of the Victorian era – foreigners, race, sexual orientation, women. Nevertheless, his theory of evolution through natural selection pointed the way forward, undermining every one of his prejudices. Moreover, before we leave Darwin with limited praise, almost *malgré lui*, Darwin himself was aware of the possibilities of his theory. This fact alone gives zest to our anticipated discussion in the *Epilogue*, about the propriety of speaking of a *Darwinian* Revolution.

Epilogue

I CIRCLE BACK TO THE QUESTION I ASKED IN MY INTRODUCTION: Was there a Darwinian Revolution? Was Darwin a great revolutionary? One thing that we certainly have seen is that Darwin was no great rebel. He drew extensively – exclusively – on the culture within which he was born and raised. Upper-middle-class British. The *Origin* shows this again and again. At the beginning of the nineteenth century, artificial selection was something much practiced as animal and plant breeders strove to meet the demands of – thanks to the Industrial Revolution – a rapidly growing urban population. Darwin came from rural Britain – Shropshire in the English Midlands – so breeding would be very familiar.

Variation again was part of this heritage. It was something with which breeders were constantly working, either using it to make new forms or struggling with the way it threatened already-existing desirable forms. Malthus and his claims about population growth and a subsequent struggle for existence were nigh commonsensical to people of Darwin's class. Natural selection was new, but not that new. Remember how this was backed by Darwin's extensive reading, including the pamphlet by Sir John Sebright, actually likening human selection to the selection brought on by the natural forces of climate and the like. The effects of natural selection, the adaptiveness of organisms, the purposes of their parts, in a country that took natural theology as its own personal property, as hymned by textbook writer Archdeacon William Paley, were as much part of the culture as the superiority of the English over all other inhabitants of Planet Earth. The division of labor was stressed by the Scottish political economist Adam Smith: "It is not from the benevolence of the butcher, the brewer, or the baker that we expect our dinner, but from their regard to their own interest." And then, the tree of life: "Genesis 2:9 And out of the ground made the LORD God to grow every tree that is pleasant to the sight, and good for food; the tree of life also in the midst of the garden, and the tree of knowledge of good and evil" (Smith 1776, 18).

No great rebel. But was Darwin revolutionary? Start with the science. Agree that many professional scientists became evolutionists after the *Origin*, but few became enthusiasts for selection. Thomas Henry Huxley: "There is no positive evidence, at present, that any group of animals has, by variation and selective breeding, given rise to another group which was, even in the least degree, infertile with the first" (Huxley 1893b, 22–79). This said, the effect of Darwin's theory on Huxley and like thinkers was massive. The aim of this new generation was to professionalize science. To take it out of the hands of the religious – Oxbridge faculty had to be ordained ministers of the established Church of England – and give it over to the secular. Huxley was typical, satisfying his intent by becoming professor (and then dean) at the new science-teaching college down in West Kensington. Huxley and fellows in their science classrooms taught little evolution and minuscule natural selection. However, they needed Darwin and the *Origin* to give their activities the odor of successful professional science, even though they did not use Darwin and his science in their activities. They basked in his status as a major, respectable scientist – major, respectable *life* scientist. They may not have wanted him in their classrooms, but they made sure that Darwin was buried in Westminster Abbey, final home of Isaac Newton.

Bowler recognizes this but downplays it as insignificant or irrelevant. Nonsense! There was something going on and Darwin was a – the – major figure. Selection aside, it was Darwin, through his consilience – something entirely missing in the writings of others – who made evolution a fact and not just (as said before) a flaky hypothesis. In any case, as we have seen, there was active science, post-*Origin*, using natural selection as its central causal tool. One refers here to those studying fast-breeding organisms: Henry Walter Bates and Fritz Müller and August Weismann, on down to E. B. Poulton and W. R. F. Weldon. And this is not to mention the amateurs or semiprofessionals such as J. W. Tutt – a headmaster as well as a founding editor of the journal *Entomologists' Record*. Here Huxley's intentions had the opposite effect. Pushing the professionalism of modern science meant that the work of amateurs was belittled. It was not held up as a paradigm of scientific activity. Today, we know better. Those amateurs were ahead of the professionals.

In the world of science in the nineteenth century, there was revolutionary science. And as one moves into and through the twentieth century – and there is every reason why one should – the status of the Darwinian Revolution multiplies without end. William Hamilton's incredibly influential discussion of hymenopteran sociality was, in major respects, simply following on Darwin's discussion in the *Origin* about how natural selection can promote sterility, so long as close relatives are benefiting. The group-selection approach of V. C.

Wynn-Edwards was cast aside as worthless and, from then on, "kin selection" and other individual-selection processes ruled supreme, popularized as they were by Richard Dawkins in *The Selfish Gene*. Few would deny that Hamilton's work was revolutionary, and if we say this of Hamilton why deny it of Darwin? It was, after all, his way of thinking that Hamilton reenergized. (Science is Heraclitean. You cannot step into the same river twice. Nothing stays still, unchanged. As we might expect, since Hamilton, there have been further developments, both empirical and theoretical. One issue has been the extent to which the queen has multiple partners. See Boomsma 2007 and 2009.)

Move now from the science to the discussions we have had in the humanities. With respect to Britain, not much of a case can be made for the revolutionary effects of Darwin's thinking in philosophy. This was true in the nineteenth century, especially if one recognizes that Social Darwinism owes little to the message of the *Origin*. The nonimpact of Darwinism continues to be basically true right through the twentieth century. In the *Oxford Handbook of the History of Analytic Philosophy*, there are four references to Darwin and over 100 to Wittgenstein. (No entries for Ruse!) It did not altogether help matters that those who turned philosophically to biology, such as J. H. Woodger, tended to do so precisely because they wanted to stake out an area that was not Darwinian. On the other hand, if we turn to America, we have the Pragmatists, so in this sense we can certainly think of Darwin's thinking as revolutionary, and we have seen that this persists into the twentieth century with such major thinkers as W. V. O. Quine and Richard Rorty. Not to mention, in the realm of education, the influence of John Dewey. We must "conceive education as the process of forming fundamental dispositions, intellectual and emotional, toward nature and fellow men" (Dewey 1916, 305).

Even if professional philosophers had trouble with Pragmatism, not to mention right-wing Catholics – "the man most responsible for destroying the moral integrity of public education in the United States" (Fr. Thomas Hickey 2017) – in the real world, Dewey's influences were lasting and profound. I quote from a recent letter I wrote to the Quaker boarding school I attended in the 1950s. I was reflecting on fifty-five years as a college professor: "The greatest message about being a teacher – it is not so much passing on knowledge – algebra and French irregular verbs – but helping young people grow into worthwhile human beings – for me, being a teacher is sacred – a privilege and an obligation." Influences, lasting and profound. Turning to moral philosophy, thinking at the level of substantive ethics, to the credit of America one can add the name of John Rawls, the leading figure in this field in the second half of the twentieth century. It is tempting to explain the Anglo-American differences in terms of the societies more generally, with America going forward into the twentieth century with the vigor that made it the dominant country of the world, and England already starting to show evidence of the decline that has ended with the tragedy of Brexit. I leave that assessment to others. Notwithstanding the authority of Ludwig Wittgenstein, that any philosopher thinks it completely irrelevant as to whether we are modified mud or modified monkeys leaves me speechless. At least one Austrian-born philosopher learnt by his mistakes. And it would be unfair to fail to note that Popper is not alone. There are today those on both sides of the Atlantic, undoubtedly still much a minority, pushing to reevaluate positively the possible relevance of Darwinism to philosophy.

Even today, most people would think that the only relationship between Darwinian evolution and Christianity is out-and-out warfare. This is indeed true of Creationism and of its offspring Intelligent Design Theory. It is true also of the New Atheists – especially those who are pushing an alternative secular religion, "humanism" or some such thing. And then there were those, such as Thomas Hardy, who were simply scared out of their minds. But from the first there were Christians – as well as those on the fringe of orthodox belief (from Unitarians to outright nonbelievers) – who saw at once its favorable implications. Against the background story of Adam and Eve and that wretched apple, Darwin's theory was a huge topic of discussion, through the nineteenth century and into the twentieth as women pushed and finally succeeded in getting the vote. We have got to stop the misguided effects of natural and sexual selections functioning inappropriately. Following St. Paul in keeping men in charge and women subservient is disastrous. From the author of *Herland*:

> The woman was deprived of the beneficent action of natural selection, and the man was then, by his own act, freed from the stern but elevating effect of sexual selection. Nothing was required of the woman by natural selection save such capacity as should please her master; nothing was required of the man by sexual selection save power to take by force, or buy, a woman. (Gilman 1915, 53)

Alas, alas: "we have perverted the order of nature, and are suffering the consequences" (Gilman 1915, 35).

Let it be emphasized that this kind of thinking, turning to the *Descent* for information and support, generated huge interest and a massive literature as the nineteenth century drew to a close and a new century, the twentieth, began. Agree or disagree, it was Darwin, Darwin, Darwin. Another fact of history that shows that claims that there was no *Darwinian* Revolution are based on a very selective reading of the literature. Or, more precisely, a nonreading of the literature.

> Regardless of whether or not readers accepted Darwin's arguments in the *Descent of Man*, all agreed that the book was a literary sensation and a must-read. Even negative reviews suggested that people read the *Descent*. In its signature lady-like tone, *Godey's Lady's Book*, the popular nineteenth-century women's magazine, noted that the book "will call forth discussion and dissent among the masterminds of the age" but demurred in conclusion, "we are not yet an avowed convert to Darwin's theories, but we find his book exceedingly interesting." The *Galaxy* proclaimed, "Whatever may be thought of Mr. Darwin's conclusions as to the origin of man, his book will be found a rich mine of facts, entertaining and curious on the highest questions of natural history." *Old and New* declared the *Descent* to be "as exciting as any novel." *Appleton's* announced that the book was the literary sensation of the month, while *Harper's* observed that "few scientific works have excited more attention" than the *Descent* as evinced by the fact that one could not open a magazine without reading about it. It appeared on prominent book lists for women's and girl's clubs until the turn of the century, and the *New York Times* reported that it was among the most popular books checked out of Manhattan public libraries as late as 1895. (Hamlin 2014, 10–11)

Either way. For and against. It is simply ludicrous to claim that there was no Darwinian Revolution. And not just on the topic of the rights of women. The same complex, shattering relationship between Christianity and evolutionary thinking, including Darwinian thinking, continued into the twentieth century. Whitehead and his creation, process theology, is a major revolutionary happening. Given that Whitehead's early work was with Bertrand Russell, he certainly would not have Darwin enthusiasts egging him on. But it is surely true that Darwin may take some pride in the way that Whitehead broke altogether new ground in his thinking about the nature of God and of His relationship to humankind.

> Maybe he wakes periodically at night,
> Wiping away the tears he doesn't know
> He has cried in his sleep, not having had time yet to tell
> Himself precisely how it is he must mourn, not having had time yet
> To elicit from his creation its invention
> Of his own solace.
>
> (Rogers 1999)

"God" is a verb, not a noun.

And so finally to literature. Here, unambiguously, Darwin did have a revolutionary effect, in fiction, in poetry, on the general public. Within a year, in a supportive context, people learnt not just about evolution but about the Darwinian mechanism of natural selection, not to mention sexual selection.

More than this. Whatever Darwin's personal views may have been, his ideas contributed to the ways in which women were starting to have a real significance for human culture. George Eliot wrestling with Darwinian themes in her writings (her final real novel *Daniel Deronda* is structured on the implications of sexual selection), Emily Dickinson on the implications of Darwinism for religious belief, and Constance Naden sufficiently secure to poke fun at them all. Darwin had changed the world. Even if one did not accept all that he said, you were supposed to know about it.

Gilbert and Sullivan were attuned to their audience (Bradley 1988). In *Princess Ida* (1884) we learn of "the apiest Ape that ever was seen." At least give him points for trying.

> He bought white ties, and he bought dress suits,
> He crammed his feet into bright tight boots
> And to start in life on a brand new plan,
> He christened himself Darwinian Man!

He is out of luck. The object of his desire

> Was a radiant Being,
> far-seeing

And apparently what she saw did not please. Women are talented:

> While a Man, however well-behaved,
> At best is only a monkey shaved!

We are in a new world. Qualify it as you will, from the start, Darwin's work was explosive with far-reaching implications. It continues to be so today. No rebel. Great revolutionary. A change in worldview indeed.

References

Abbott, P. et al., 2011. Inclusive fitness theory and eusociality. *Nature* 471: E1–4.

Adovasio, J. M., O. Soffer, and J. Page. 2007. *The Invisible Sex: Uncovering the True Roles of Women in Prehistory*. New York: Collins.

Agassiz, E. C., Editor. 1885. *Louis Agassiz: His Life and Correspondence*. Boston: Houghton Mifflin.

Agassiz, L. 1833–1943. *Recherches sur les Poissons Fossiles*. Neuchâtel: Imprimerie de Petitpierre.

1859. *Essay on Classification*. London: Longman, Brown, Green, Longmans, and Roberts.

Allen, E., et al. 1975. Letter to the editor. *New York Review of Books*, sec. 22, 18, pp. 43–44.

1976. Sociobiology: a new biological determinism. *BioScience* 26: 182–86.

Allen, G. E. 1978. *Thomas Hunt Morgan: The Man and His Science*. Princeton: Princeton University Press.

Allison, A. C. 1954a. Protection by the sickle-cell trait against subtertian malarial infection. *British Medical Journal* 1: 290.

1954b. The distribution of the sickle-cell trait in East Africa and elsewhere and its apparent relationship to the incidence of subtertian malaria. *Transactions of the Royal Society of Tropical Medical Hygiene* 48: 312.

Anon. 1860a. Natural selection. *All the Year Round* 3 (63): 293–99.

1860b. Species. *All the Year Round* 3 (58): 174–78.

1861. Transmutation of species. *All the Year Round* 4 (98): 519–21.

Anselm, St. 1903. *Anselm: Proslogium, Monologium, an Appendix on Behalf of the Fool by Gaunilon; and Cur Deus Homo*. S. N. Deane. Chicago: Open Court.

Appleman, P. 1996. *New and Selected Poems, 1956–1996*. Fayetteville: University of Arkansas Press.

2014. *The Labyrinth: God, Darwin, and the Meaning of Life*. New York: Quantuck Lane Press.

Aquinas, St. T. 1947. *Compendium Theologiae*. St. Louis and London: Herder.

1952. *Summa Theologica, I*. London: Burns, Oates and Washbourne.

Arthur, W. 2021. *Understanding Evo-Devo*. Cambridge: Cambridge University Press.

Augustine. 1991. *On Genesis*. Translator R. J. Teske. Washington, DC: Catholic University of America Press.

[396] 1998. *Confessions*. Translator H. Chadwick. Oxford: Oxford University Press.

[413–426] 2000. *The City of God against the Pagans*. Translator M. Dons. New York: Random House.

Ayer, A. J. 1936. *Language, Truth and Logic*. London: Gollancz.

Baldwin, T. 1990. *G. E. Moore*. London: Routledge & Kegan Paul.

Barbour, I. 1997. *Religion and Science: Historical and Contemporary Issues*. San Francisco: Harper.

Barnes, J., Editor. 1984. *The Complete Works of Aristotle*. Princeton, NJ: Princeton University Press.

Barrie, J. M. 1908. *What Every Woman Knows* (Play).

Bartholomew, M. 1973. Lyell and evolution: an account of Lyell's response to the prospect of an evolutionary ancestry for man. *British Journal for the History of Science* 6: 261–303.

Bates, H. W. 1862. Contributions to an insect fauna of the Amazon Valley. *Transactions of the Linnean Society of London* 23: 495–515.

Bateson, W. 1894. *Materials for the Study of Variation, Treated with Especial Regard to Discontinuity in the Origin of Species*. London: Macmillan.

1902. *Mendel's Principles of Heredity: A Defence, with a Translation of Mendel's Original Papers on Hybridisation*. Cambridge: Cambridge University Press.

Baynes, T. S. 1873. Darwin on expression. *Edinburgh Review* 137: 492–508.

Behe, M. 1996. *Darwin's Black Box: The Biochemical Challenge to Evolution*. New York: Free Press.

Bell, A., and S. Weinberg. 1978. *Homosexualities – A Study of Diversity among Men and Women*. New York: Simon & Schuster.

Bergson, H. 1907. *L'évolution créatrice*. Paris: Alcan.

1911. *Creative Evolution*. New York: Holt.

Biological Sciences Curriculum Study. 1963. *Biological Science: Molecules to Man*. Boston: Houghton Mifflin.

Blumenthal, P. 2023. How Mike Johnson Helped Open the Door to Creationism in Louisiana Public Schools. *Huffington Post*, December 10. www.huffpost.com/entry/mike-johnson-creationism-schools_n_657380ffe4b09724b4342739.

Boomsma, J. 2007. Kin selection versus sexual selection: why the ends do not meet. *Current Biology* 17: R673–83.

2009. Lifetime monogamy and the evolution of eusociality. *Philosophical Transactions of the Royal Society of London. Series B, Biological Sciences* 364: 3191–208.

Bowler, P. J. 1983. *The Eclipse of Darwinism: Anti-Darwinism Evolution Theories in the Decades around 1900*. Baltimore, MD: Johns Hopkins University Press.

1988. *The Non-Darwinian Revolution: Reinterpreting a Historical Myth*. Baltimore, MD: Johns Hopkins University Press.

1989. *The Mendelian Revolution: The Emergence of Hereditarian Concepts in Modern Science and Society*. London: The Athlone Press.

2013. *Darwin Deleted: Imagining a World without Darwin*. Chicago: University of Chicago Press.

Boyle, R. [1688] 1966. A disquisition about the final causes of natural things. *The Works of Robert Boyle*. Editor T. Birch, 5: 392–444. Hildesheim: Georg Olms.

1996. *A Free Enquiry into the Vulgarly Received Notion of Nature*. Editors E. B. Davis, and M. Hunter. Cambridge: Cambridge University Press.

Bradley, I. C. 1988. *The Annotated Gilbert and Sullivan: Trial by Jury, The Sorcerer, Patience, Princess Ida, Ruddigore, The Yeomen of the Guard*. London: Penguin.

Brantley, R. E. 2014. The interrogative mood of Emily Dickinson's quarrel with God. *Religion & Literature*, 46, 157–65.
Brewster, D. 1854. *More Worlds than One: The Creed of the Philosopher and the Hope of the Christian*. London: Camden Hotten.
Broad, C. D. 1944. Critical notice of Julian Huxley's *Evolutionary Ethics*. *Mind* 53: 344–67.
Brooks, R. 2021. Darwin's closet: the queer sides of the descent of man (1871). *Zoological Journal of the Linnean Society* 191 (2): 323–46.
Browne, J. 1995. *Charles Darwin: Voyaging. Volume 1 of a Biography*. London: Jonathan Cape.
2002. *Charles Darwin: The Power of Place. Volume 2 of a Biography*. London: Jonathan Cape.
Burchfield, J. D. 1975. *Lord Kelvin and the Age of the Earth*. New York: Science History Publications.
Burkeman, O. 2021. The clockwork universe: is freewill an illusion? *Guardian*, April 27.
Burkhardt, R. W. 2005. *Patterns of Behavior: Konrad Lorenz, Niko Tinbergen, and the Founding of Ethology*. Chicago: University Chicago Press.
Calvin, J. 1536. *Institutes of the Christian Religion*. Grand Rapids: Eerdmans.
Carnegie, A. 1889. The Gospel of Wealth. *North American Review* 148: 653–65.
Ceci, S. J., and W. M. Williams. 2009. *The Mathematics of Sex: How Biology and Society Conspire to Limit Talented Women and Girls*. New York: Oxford University Press.
Chambers, R. 1844. *Vestiges of the Natural History of Creation*. London: Churchill.
Coleman, W. 1964. *Georges Cuvier Zoologist: A Study in the History of Evolution Theory*. Cambridge, MA: Harvard University Press.
Conan Doyle, A. 1890. *The Sign of the Four*. London: Nelson.
Cooper, J. M., Editor. 1997. *Plato: Complete Works*. Indianapolis: Hackett.
Coyne, J. A. 2015. *Faith Versus Fact: Why Science and Religion Are Incompatible*. New York: Viking.
2017. More dumb claims that environmental epigenetics will completely revise our view of evolution. *Why Evolution Is True Blog*. https://whyevolutionistrue.com/2017/10/27/more-dumb-theorizing-that-epigenetics-as-lamarckian-inheritance-will-completely-revise-our-view-of-evolution/.
2019. Epigenetics: The return of Lamarck? Not so fast! *Why Evolution Is True Blog*. https://whyevolutionistrue.com/2018/08/26/epigenetics-the-return-of-lamarck-not-so-fast/.
Coyne, J. A., and H. A. Orr. 2004. *Speciation*. Sunderland, MA: Sinauer.
Cuvier, G. 1817. *Le règne animal distribué d'aprés son organisation, pour servir de base à l'histoire naturelle des animaux et d'introduction à l'anatomie comparée*. Paris: Déterville.
Darwin, C. 1839. Observations on the parallel roads of Glen Roy, and of other parts of Lochaber in Scotland, with an attempt to prove that they are of marine origin. *Philosophical Transactions of the Royal Society of London* 129: 39–81.
1842. *The Structure and Distribution of Coral Reefs: Being the First Part of the Geology of the Voyage of the Beagle, under the Command of Capt. Fitzroy, R.N. during the Years 1832 to 1836*. London: Smith Elder.
1845. *Journal of Researches into the Natural History and Geology of the Countries Visited during the Voyage of H.M.S. Beagle Round the World [2nd ed.]*. London: John Murray.

1851a. *A Monograph of the Fossil Lepadidae; or, Pedunculated Cirripedes of Great Britain*. London: Palaeontographical Society.
1851b. *A Monograph of the Sub-Class Cirripedia, with Figures of all the Species: The Lepadidae; or Pedunculated Cirripedes*. London: Ray Society.
1854a. *A Monograph of the Fossil Balanidae and Verrucidae of Great Britain*. London: Palaeontographical Society.
1854b. *A Monograph of the Sub-Class Cirripedia, with Figures of all the Species: The Balanidge (or Sessile Cirripedes); the Verrucidae, and C*. London: Ray Society.
1859. *On the Origin of Species by Means of Natural Selection, or the Preservation of Favoured Races in the Struggle for Life*. London: John Murray.
1861. *Origin of Species, Third Edition*. London: John Murray.
1862. *On the Various Contrivances by which British and Foreign Orchids are Fertilized by Insects, and On the Good Effects of Intercrossing*. London: John Murray.
1868. *The Variation of Animals and Plants under Domestication*. London: John Murray.
1869. *On the Origin of Species*. 5th ed. London: John Murray.
1871. *The Descent of Man and Selection in Relation to Sex*. London: John Murray.
1958. *The Autobiography of Charles Darwin 1809–1882: With the Original Omissions Restored. Edited and with Appendix and Notes by His Grand-Daughter Nora Barlow*. London: Collins.
1975. *Charles Darwin's Natural Selection, Being the Second Part of His Big Species Book Written from 1856 to 1858*. Editor R. C. Stauffer. Cambridge: University of Cambridge Press.
1985–. *The Correspondence of Charles Darwin*. Cambridge: Cambridge University Press.
1987. *Charles Darwin's Notebooks, 1836–1844*. Editors P. H. Barrett, P. J. Gautrey, S. Herbert, D. Kohn, and S. Smith. Ithaca, NY: Cornell University Press.
Darwin, E. 1789. *The Botanic Garden (Part II, The Loves of the Plants)*. London: J. Johnson.
[1794–1796] 1801. *Zoonomia; or, The Laws of Organic Life*. 3rd ed. London: J. Johnson.
1803. *The Temple of Nature*. London: J. Johnson.
Darwin, F. 1887. *The Life and Letters of Charles Darwin, Including an Autobiographical Chapter*. London: John Murray.
1909. *The Foundations of the Origin of Species: Two Essays Written in 1842 and 1844*. Cambridge: Cambridge University Press.
Dawkins, R. 1976. *The Selfish Gene*. Oxford: Oxford University Press.
1983. Universal Darwinism. *Evolution from Molecules to Men*. Editor D. S. Bendall, 403–25. Cambridge: Cambridge University Press.
1986. *The Blind Watchmaker*. New York: Norton.
2006. *The God Delusion*. New York: Houghton, Mifflin, Harcourt.
Dembski, W. A. 1998. *The Design Inference: Eliminating Chance through Small Probabilities*. Cambridge: Cambridge University Press.
Dennett, D. C. 1995. *Darwin's Dangerous Idea*. New York: Simon & Schuster.
Depew, D. 2023. Richard Lewontin and Theodosius Dobzhansky: Genetics, race, and the anxiety of influence. *Biological Theory*. https://link.springer.com/article/10.1007/s13752-023-00452-2.
DeSalle, R., and I. Tattersall. 2022. *Understanding Race*. Cambridge: Cambridge University Press.

Descartes, R. 1985. *The Philosophical Writings, Volume I.* Translators J. Cottingham, R. Stoothoff, and D. Murdoch. Cambridge: Cambridge University Press.

Desmond, A. 1998. *Huxley: From Devil's Disciple to Evolution's High Priest.* London: Penguin.

Desmond, A., and J. Moore. 2009. *Darwin's Sacred Cause: How a Hatred of Slavery Shaped Darwin's Views on Human Evolution.* New York: Houghton Mifflin Harcourt.

Dewey, J. 1897. What education is. *School Journal* LIV, 3, 77–80.

1906. The experimental theory of knowledge. *Mind* 15: 293–307.

1910. *How We Think.* Boston: D. C. Heath.

1916. *Democracy and Education.* New York: Macmillan.

1976. *The Collected Works of John Dewey.* Carbondale: Southern Illinois University Press.

Di Gregorio, M., and N. W. Gill, Editors. 1990. *Charles Darwin's Marginalia, Volume 1.* New York: Garland.

Dias, E., and R. Graham. 16 June 2023. At the Southern Baptist Convention, a Call to Enforce Biblical Gender Roles. *New York Times,* sec. A, p. 16.

Dickens, C. [1860] 1948. *Great Expectations.* London: Oxford University Press.

[1865] 1948. *Our Mutual Friend.* Oxford: Oxford University Press.

Dickinson, E. 1960. *The Complete Poems of Emily Dickinson.* New York: Little, Brown.

Diderot, D. 1943. *Diderot: Interpreter of Nature.* New York: International Publishers.

[1796] 1972. *The Nun.* London: Penguin.

Dijksterhuis, E. J. 1961. *The Mechanization of the World Picture.* Oxford: Oxford University Press.

Dixon, M., and G. Radick. 2009. *Darwin in Ilkley.* Stroud, UK: History Press.

Dobzhansky, T. 1937. *Genetics and the Origin of Species.* New York: Columbia University Press.

1943. Temporal changes in the composition of populations of Drosophila pseudoobscura in different environments. *Genetics* 28: 162–86.

1951. *Genetics and the Origin of Species (Third Edition).* New York: Columbia University Press.

1962. *Mankind Evolving.* New Haven, CT: Yale University Press.

Draper, J. W. 1875. *History of the Conflict between Religion and Science.* New York: Appleton.

Duncan, D., Editor. 1908. *Life and Letters of Herbert Spencer.* London: Williams and Norgate.

Dupré, J., Editor. 1987. *The Latest on the Best: Essays on Evolution and Optimality.* Cambridge, MA: MIT Press.

2012. *Processes of Life: Essays in the Philosophy of Biology.* Oxford: Oxford University Press.

Dyble, M., G. D. Salali, N. Chaudhary, A. Page, D. Smith, J. Thompson, L. Vinicius, R. Mace, and A. B. Migliano. 2015. Sex equality can explain the unique social structure of hunter-gatherer bands. *Science* 348 (6236): 796–98.

Eldredge, N., and S. J. Gould. 1972. Punctuated equilibria: an alternative to phyletic gradualism. *Models in Paleobiology.* Editor T. J. M. Schopf, 82–115. San Francisco: Freeman, Cooper.

Evans, M. A., and H. E. Evans. 1970. *William Morton Wheeler, Biologist.* Cambridge, MA: Harvard University Press.

Fairbanks, D. J. 2020. Mendel and Darwin: untangling a persistent enigma. *Heredity* 124, 263–73.
Farley, J. 1977. *The Spontaneous Generation Controversy from Descartes to Oparin.* Baltimore, MD: Johns Hopkins University Press.
Farlow, J. O., C. V. Thompson, and D. E. Rosner. 1976. Plates of the dinosaur Stegosaurus: forced convection heat loss fins? *Science* 192: 1123–25.
Fergusson, D. 2009. Darwin and providence. *Theology after Darwin.* Editors M. Northcott, and R. J. Berry, 73–88. Carlisle: Paternoster Press.
Fisher, R. A. 1930. *The Genetical Theory of Natural Selection.* Oxford: Oxford University Press.
Ford, E. B. 1931. *Mendelism and Evolution.* London: Methuen.
1964. *Ecological Genetics.* London: Methuen.
Francis, M. 2007. *Herbert Spencer and the Invention of Modern Life.* Ithaca, NY: Cornell University Press.
Freud, S. 1935. Letter to an American Mother. *Gay/Lesbian Resources.* http://psychpage.com/gay/library/freudsletter.html.
1960. *Letters of Sigmund Freud, 1873–1939.* Translators and editors T. Stern, and J. Stern. New York: Basic Books.
Fry, D. P. 2013. *War, Peace, and Human Nature: The Convergence of Evolutionary and Cultural Views.* Editor D. P. Fry, Oxford: Oxford University Press.
Gamble, E. B. 1894. *The Evolution of Woman: An Inquiry into the Dogma of Her Inferiority to Man.* New York: Putnam.
Gare, A. 2002. The roots of postmodernism: Schelling, process philosophy and post-structuralism. *Process and Difference: Between Cosmological and Poststructuralist Postmodernisms.* Editors C. Keller, and A. Daniell, 31–54. Albany: SUNY Press.
Gayon, J. 2013. Darwin and Darwinism in France after 1900. *The Cambridge Encyclopedia of Darwin and Evolutionary Thought.* Editor M. Ruse, 300–12. Cambridge: Cambridge University Press.
Gibson, A. 2013. Edward O. Wilson and the organicist tradition. *Journal of the History of Biology* 46: 599–630.
Gibson, M. 2023. Chariots of philosophical fire. *City Journal (Manhattan Institute)*: www.city-journal.org/article/chariots-of-philosophical-fire.
Gilman, C. P. [1915] 1979. *Herland.* New York: Pantheon.
Gish, D. 1973. *Evolution: The Fossils Say No!* San Diego: Creation-Life.
Gissing, G. [1891] 1976. *New Grub Street.* London: Penguin.
Gladwin, M. T., G. Kato, and E.M. Novelli. 2021. *Sickle Cell Disease.* New York: McGraw Hill.
Goethe, J. W. [1790] 1946. On the metamorphosis of plants. *Goethe's Botany, Chronica Botanica.* Editor A. Arber, 63–126. Gotha: Ettinger.
Gould, S. J. 1980. Is a new and general theory of evolution emerging? *Paleobiology* 6: 119–30.
1981. *The Mismeasure of Man.* New York: Norton.
Gould, S. J., and N. Eldredge. 1977. Punctuated equilibria: the tempo and mode of evolution reconsidered. *Paleobiology* 3: 115–51.
Gould, S. J., and R. C. Lewontin. 1979. The spandrels of San Marco and the Panglossian paradigm: a critique of the adaptationist programme. *Proceedings of the Royal Society of London, Series B: Biological Sciences* 205: 581–98.

Graham, W. 1881. *The Creed of Science: Religious, Moral, and Social*. London: Kegan Paul.

Grant, B. R., and P. R. Grant. 1993. Evolution of Darwin's Finches caused by a rare climatic event. *Proceedings of the Royal Society: Biological Sciences* 251: 111–17.

2003. What Darwin's Finches can teach us about the evolutionary origin and regulation of biodiversity. *BioScience* 53: 965–75.

Grant, P. R., and B. R. Grant. 1995. Predicting microevolutionary responses to directional selection on heritable variation. *Evolution* 49: 241–51.

Grant, B. S. 2021. *Observing Evolution: Peppered Moths and the Discovery of Parallel Melanism*. Baltimore, MD: Johns Hopkins University Press.

Green, T. H. 1883. *Prolegomena to Ethics*. A. C. Bradley (ed.), Oxford: Clarendon Press.

Greene, G. [1973] 1974. *The Honorary Consul*. New York: Simon & Schuster.

Greene, J. C., and M. Ruse 1996. On the nature of the evolutionary process: the correspondence between Theodosius Dobzhansky and John C. Greene. *Biology and Philosophy* 11: 445–91.

Gunn, J. A. 1920. *Bergson and His Philosophy*. New York: E. P. Dutton.

Haeckel, E. 1866. *Generelle Morphologie der Organismen*. Berlin: Georg Reimer.

[1868] 1876. *The History of Creation*. Translator E. Ray Lankester. London: Kegan Paul.

Haldane, J. B. S. 1932. *The Causes of Evolution*. New York: Cornell University Press.

Hall, A. R. 1954. *The Scientific Revolution 1500–1800: The Formation of the Modern Scientific Attitude*. London: Longman, Green and Company.

Hamilton, W. D. 1964. The genetical evolution of social behaviour. *Journal of Theoretical Biology* 7: 1–32.

Hamlin, K. A. 2014. *From Eve to Evolution: Darwin, Science, and Women's Rights in Gilded Age America*. Chicago: University of Chicago Press.

Harrington, A. 1996. *Reenchanted Science: Holism in German Culture from Wilhelm II to Hitler*. Princeton, NJ: Princeton University Press.

Harris, M., and J. Johnson, Editors. 1998. *The Journals of George Eliot*. Cambridge: Cambridge University Press.

Henderson, L. J. 1917. *The Order of Nature*. Cambridge, MA: Harvard University Press.

Herschel, J. F. W. 1830. *Preliminary Discourse on the Study of Natural Philosophy*. London: Longman, Rees, Orme, Brown, Green, and Longman.

1841. Review of Whewell's history and philosophy. *Quarterly Review* 135: 177–238.

Hickey, T. 2017. "Thomistic evolution": development of doctrine or diabolical deception? https://kolbecenter.org/thomistic-evolution-development-doctrine-diabolical-deception/.

Hobhouse, L. T. 1913. *Development and Evolution: An Essay towards a Philosophy of Evolution*. London: Macmillan.

Honenberger, P. 2018. Darwin among the philosophers: Hull and Ruse on Darwin, Herschel, and Whewell. *HOPOS: The Journal of the International Society for the History of Philosophy of Science* 8: 278–309.

Hoquet, T. 2018. *Revisiting the Origin of Species: The Other Darwins (History and Philosophy of Biology)*. London: Routledge.

Hrdy, S. B. 1999. *Mother Nature: A History of Mothers, Infants, and Natural Selection*. New York: Pantheon Books.

Hull, D. 1969. What the philosophy of biology is not. *Synthese* 20: 57–184.

Hume, D. [1739–40] 1978. *A Treatise of Human Nature.* Oxford: Oxford University Press.

Huxley, J. S. 1912. *The Individual in the Animal Kingdom.* Cambridge: Cambridge University Press.

1942. *Evolution: The Modern Synthesis.* London: Allen & Unwin.

1943. *Evolutionary Ethics.* Oxford: Oxford University Press.

Huxley, L. 1900. *The Life and Letters of Thomas Henry Huxley.* London: Macmillan.

Huxley, T. H. 1877. *American Addresses, with a Lecture on the Study of Biology.* London: Macmillan.

[1859] 1893. The Darwinian hypothesis. *Collected Essays: Darwiniana.* Editor T. H. Huxley, 1–21. London: Macmillan.

1863. *Evidence as to Man's Place in Nature.* London: Williams and Norgate.

1880. The coming of age of "The origin of species." *Science* 1: 15–20.

1893a. *Evolution and Ethics.* London: Macmillan.

[1860] 1893b. The origin of species. *Darwiniana.* Editor T. H. Huxley, 22–79. London: Macmillan.

1900. *Discourses: Biological and Geological.* New York and London: Appleton.

Hyatt, A. 1897. Cycle in the life of the individual (ontogeny) and in the evolution of its own group (phylogeny). *Proceedings of the American Academy of Arts and Sciences* 32: 209–24.

Jablonski, N. G., and G. Chaplin. 2017. The colours of humanity: the evolution of pigmentation in the human lineage. *Philosophical Transactions: Biological Sciences* 372 (1724): 1–8.

James, H. 1873. Middlemarch. *Galaxy* 15: 424–28.

James, W. 1907. *Pragmatism: A New Name for Some Old Ways of Thinking.* New York: Longmans, Green.

Jarvenpa, R., and H. J. Brumbach. 2014. Hunter-gatherer gender and identity. *The Oxford Handbook of the Archaeology and Anthropology of Hunter-Gatherers.* Editors V. Cummings, P. Jordan, and M. Zvelebil, 1243–65. Oxford: Oxford University Press.

Jenkin, F. 1867. Review of "The origin of species." *The North British Review* 46: 277–318.

Johnson, P. E. 1991. *Darwin on Trial.* Washington, DC: Regnery Gateway.

Kallmann, F. 1952. Comparative twin study on the genetic aspects of male homosexuality. *The Journal of Nervous and Mental Disease* 115: 284–98.

Kant, I. [1790] 2000. *Critique of the Power of Judgment.* Editor P. Guyer. Cambridge: Cambridge University Press.

Kavaloski, V. 1974. The "vera causa" principle: an historico-philosophical study of a metatheoretical concept from Newton through Darwin. PhD dissertation, University of Chicago.

Kellogg, V. L. 1905. *Darwinism Today: A Discussion of the Present-Day Scientific Criticism of the Darwinian Selection Theories, Together with a Brief Account of the Principle Other Proposed Auxiliary and Alternative Theories of Species-Forming.* New York: Henry Holt.

Kingsley, C. 1863. *The Water-Babies: A Fairy Tale for a Land-Baby.* London: Macmillan.

Kipling, R. 2001. *Collected Poems.* Ware: Wordsworth Poetry Library.

Kruse, F. E. 2010. Peirce, God, and the "Transcendentalist Virus." *Transactions of the Charles S. Peirce Society* 46: 386–400.

Kuhn, T. 1962. *The Structure of Scientific Revolutions*. Chicago: University of Chicago Press.

Laland, K. N., T. Uller, M. W. Feldman, K. Sterelny, B. Müller, A. Moczek, E. Jablonka, and J. Odling-Smee. 2015. The extended evolutionary synthesis: its structure, assumptions and predictions. *Proceedings of the Royal Society, B* https://kevintshoemaker.github.io/EECB-703/Laland%20et%20al.%20-%202015%20-%20The%20extended%20evolutionary%20synthesis%20its%20structure.pdf.

Lamarck, J. B. 1809. *Philosophie Zoologique*. Paris: Dentu.

Larson, E. J. 1997. *Summer for the Gods: The Scopes Trial and America's Continuing Debate over Science and Religion*. New York: Basic Books.

Lawrence, D. H. [1915] 1949. *The Rainbow*. London: Penguin.

LeDrew, S. 2016. *The Evolution of Atheism: The Politics of a Modern Movement*. Oxford: Oxford University Press.

Lennox, J. G. 1993. Darwin *was* a teleologist. *Biology and Philosophy* 8: 409–21.

Le Page, M. 2017. Blind cave fish lost eyes by unexpected evolutionary process. *New Scientist*, August 12.

LeVay, S. 2010. *Gay, Straight, and the Reason Why: The Science of Sexual Orientation*. Oxford: Oxford University Press.

Lewontin, R. C. 1974. *The Genetic Basis of Evolutionary Change*. New York: Columbia University Press.

1976. Sociobiology – a caricature of Darwinism. *PSA: Proceedings of the Biennial Meeting of the Philosophy of Science Association* 2: 22–31.

1995. Dobzhansky – theoretician without tools. *Genetics of Natural Populations: The Continuing Importance of Theodosius Dobzhansky*. Editor L. Levine, 87–101. New York: Columbia University Press.

Lipscomb, B. J. B. 2021. *The Women Are Up to Something: How Elizabeth Anscombe, Philippa Foot, Mary Midgley, and Iris Murdoch Revolutionized Ethics*. Oxford: Oxford University Press.

Lorenz, K. 1966. *On Aggression*. London: Methuen.

Lucas, J. R. 1979. Wilberforce and Huxley: a legendary encounter. *Historical Journal* 22: 313–30.

Lucretius. 1950. *Of the Nature of Things*. Translator W. E. Leonard. London: Dutton (Everyman's Library).

Luther, M. 1914. Martin Luther's Last Sermon in Wittenberg … Second Sunday in Epiphany, 17 January 1546. *Dr. Martin Luthers Werke: Kritische Gesamtausgabe*. Editor M. Luther, 7. Vol. 51. Weimar: Herman Boehlaus Nachfolger.

Lyell, C. 1830–1833. *Principles of Geology: Being an Attempt to Explain the Former Changes in the Earth's Surface by Reference to Causes Now in Operation*. London: John Murray.

Lyell, K. 1881. *Life, Letters and Journals of Sir Charles Lyell, Bart*. London: John Murray.

Majerus, M. E. N. 1998. *Melanism: Evolution in Action*. Oxford: Oxford University Press.

Malthus, T. R. [1826] 1914. *An Essay on the Principle of Population (Sixth Edition)*. London: Everyman.

Marsden, G. M. 1980. *Fundamentalism and American Culture: The Shaping of Twentieth Century Evangelicalism 1870–1925*. Oxford: Oxford University Press.

McEwan, I. 1997. *Enduring Love*. London: Cape.

[2005] 2006. *Saturday*. London: Vintage.

McMullin, E. 1983. Values in Science. *PSA 1982*. Editors P. D. Asquith, and T. Nickles, 3–28. East Lansing, MI: Philosophy of Science Association.
McPherson, J. M. 1988. *Battle Cry of Freedom: The Civil War Era*. New York: Oxford University Press.
Medawar, P. B. 1961. Review of the phenomenon of man. *Mind* 70: 99–106.
Members of the Johns Hopkins University. 1883. *Studies in Logic*. Boston, MA: Little, Brown, and Company.
Mervosh, S. 2023. DeSantis faces swell of criticism over Florida's new standards for black history. *New York Times*, July 23, A, 13.
Mill, J. S. 1862. *System of Logic*, 5th ed. New York: Harper.
Mivart, S. G. J. 1874. [Review] Researches into the early history of mankind [etc.]. *Quarterly Review* 137: 40–77.
Moore, A. 1889. The Christian doctrine of God. *Lux Mundi*. Editor C. Gore, 57–109. London: John Murray.
Moore, G. E. 1903. *Principia Ethica*. Cambridge: Cambridge University Press.
Moore, J. 1979. *The Post-Darwinian Controversies: A Study of the Protestant Struggle to Come to Terms with Darwin in Great Britain and America, 1870–1900*. Cambridge: Cambridge University Press.
Müller, F. 1879. Ituna and Thyridia: a remarkable case of mimicry in butterflies. *Transactions of the Entomological Society of London*, xx–xxix.
Murphey, M. G. 1968. Kant's children: the Cambridge pragmatists. *Transactions of the Charles S. Peirce Society* 4: 3–33.
Museum of Paleontology. 2023. *Evolution 101*. Berkeley: University of California.
Naden, C. 1999. *Poetical Works of Constance Naden*. Kernville, CA: High Sierra Books.
Newman, J. H. 1973. *The Letters and Diaries of John Henry Newman, XXV*. Editors C. S. Dessain, and T. Gornall. Oxford: Clarendon Press.
Nicholson, D., and R. Gawne. 2014. Rethinking Woodger's legacy in the philosophy of biology. *Journal of the History of Biology* 47 (2): 243–92.
Noll, M. 2002. *America's God: From Jonathan Edwards to Abraham Lincoln*. New York: Oxford University Press.
Numbers, R. L. 1992. *Prophetess of Health: Ellen G. White and the Origins of Seventh-day Adventist Health Reform (Revised Edition)*. Knoxville: University of Tennessee Press.
2006. *The Creationists: From Scientific Creationism to Intelligent Design*. Standard ed. Cambridge, MA: Harvard University Press.
Obama, B., and M. Robinson. 2015. President Obama and Marilynne Robinson: A Conversation in Iowa I. *New York Review of Books*, November 5.
O'Connell, J., and M. Ruse. 2021. *Social Darwinism (Cambridge Elements on the Philosophy of Biology)*. Cambridge: Cambridge University Press.
Ostrer, H., and K. Skoreck. 2013. The population genetics of the Jewish people. *Human Genetics* 132: 119–27.
Owen, R. 1849. *On the Nature of Limbs*. London: Voorst.
1860. *Paleontology or A Systematic Summary of Extinct Animals and Their Geological Relations*. Edinburgh: Adam and Charles Black.
Paley, W. [1802] 1819. *Natural Theology (Collected Works: IV)*. London: Rivington.
Pearson, K. 1900. *The Grammar of Science*. 2d ed. London: Black.
Peel, J. D. Y. 1971. *Herbert Spencer: The Evolution of a Sociologist*. London: Heinemann.

Peirce, C. S. 1877. The fixation of belief. *Popular Science Monthly* 12: 1–15.
1878. How to make our ideas clear. *Popular Science Monthly* 2: 286–302.
1883. A theory of probable inference. *Studies in Logic by Members of the Johns Hopkins University*. Editor C. S. Peirce, 126–81. Boston: Little, Brown and Co.
[1893] 1935. Evolutionary love (*Monist*, 3, 176–200). *Collected Papers of C.S. Peirce*. Edited by C. Hartshorne, and P. Weiss, 287–317. Cambridge, MA: Belknap Press of Harvard University.
1958. Conclusion of the history of science lectures [Lowell lectures on the history of science, 1892]. *Values in a World of Chance: Selected Writings of Charles S. Peirce (1839–1914)*. Editor P. P. Wiener, 257–60. Garden City, NY: Doubleday.
Plantinga, A. 1993. *Warrant and Proper Function*. New York: Oxford University Press.
Popper, K. R. 1959. *The Logic of Scientific Discovery*. New York: Basic Books.
1974. Darwinism as a metaphysical research programme. *The Philosophy of Karl Popper*. Editor P. A. Schilpp, 133–43. Vol. 1. LaSalle, IL: Open Court.
Poulton, E. B. 1890. *The Colours of Animals*. London: Kegan Paul, Trench, Truebner.
1908. *Essays on Evolution, 1889–1907*. Oxford: Oxford University Press.
Powell, B. 1855. *Essays on the Spirit of the Inductive Philosophy*. London: Longman, Brown, Green, and Longmans.
1860. On the study of the evidences of Christianity. *Essays and Reviews*. Editor J. W. Parker, 94–144. London: Longman, Green, Longman, and Roberts.
Provine, W. B. 1971. *The Origins of Theoretical Population Genetics*. Chicago: University of Chicago Press.
1986. *Sewall Wright and Evolutionary Biology*. Chicago: University of Chicago Press.
Quine, W. V. O. 1969. *Ontological Relativity and Other Essays*. New York: Columbia University Press.
Radick, G. 2023. *Disputed Inheritance: The Battle over Mendel and the Future of Biology*. Chicago: University of Chicago Press.
Rawls, J. 1971. *A Theory of Justice*. Cambridge, MA: Harvard University Press.
Reich, D. 2018. *Who We Are and How We Got Here: Ancient DNA and the New Science of the Human Race*. New York: Pantheon.
Reiss, M., and M. Ruse. 2023. *The New Biology: The Battle between Mechanism and Organicism*. Cambridge, MA: Harvard University Press.
Rhees, R., Editor. 1981. *Ludwig Wittgenstein: Personal Recollections*. Oxford: Blackwell.
Richards, E. 2017. *Darwin and the Making of Sexual Selection*. Chicago: University of Chicago Press.
Richards, R. J. 2002. *The Romantic Conception of Life: Science and Philosophy in the Age of Goethe*. Chicago: University of Chicago Press.
2008. *The Tragic Sense of Life: Ernst Haeckel and the Struggle over Evolutionary Thought*. Chicago: University of Chicago Press.
Roberts, S. 1993. *Sophia Jex-Blake: A Woman Pioneer in Nineteenth-Century Medical Reform*. London: Routledge.
Robinson, M. [1998] 2005. *The Death of Adam: Essays on Modern Thought*. New York: Picador.
[2004] 2005. *Gilead*. London: Virago.
2008. *Home*. New York: Farrar, Straus, and Giroux.
2014. *Lila*. London: Virago.
Robson, G. C. and O. W. Richards. 1936. *The Variation of Animals in Nature*. London: Longmans, Green & Co.

Rogers, P. 1999. *The Dream of the March Wren*. Minneapolis, MN: Milkweed Editions.
2001. *Song of the World Becoming: New and Collected Poems 1981–2001*. Minneapolis: Milkweed.
Rorty, R. 1998. *Truth and Progress: Philosophical Papers, III*. Cambridge: Cambridge University Press.
Rosenberg, N. A., J. K. Pritchard, J. L. Weber, H. M. Cann, K. K. Kidd, L. A. Zhivotovsky, and M. Feldman. 2002. Genetic structure of human populations. *Science* 298: 2381–85.
Rowell, G. 1974. *Hell and the Victorians: A Study of the Nineteenth-Century Theological Controversies Concerning Eternal Punishment and the Future Life*. Oxford: Oxford University Press.
Rudolph, J. L. 2015. That the Soviet launch of *Sputnik* caused the revamping of American science education. *Newton's Apple and Other Myths about Science*. Editors R. L. Numbers, and K. Kampourakis, 186–92. Cambridge, MA: Harvard University Press.
Rudwick, M. J. S. 1970. The strategy of Lyell's Principles of Geology. *Isis* 61: 5–33.
Rupik, G. 2024. *Remapping Biology with Goethe, Schelling, and Herder: Romanticizing Evolution*. London: Routledge.
Ruse, M. 1973. *The Philosophy of Biology*. London: Hutchinson.
1975a. Charles Darwin and artificial selection. *Journal of the History of Ideas* 36: 339–50.
1975b. Darwin's debt to philosophy: an examination of the influence of the philosophical ideas of John F.W. Herschel and William Whewell on the development of Charles Darwin's theory of evolution. *Studies in History and Philosophy of Science* 6: 159–81.
1977. Karl Popper's philosophy of biology. *Philosophy of Science* 44: 638–61.
1979. *The Darwinian Revolution: Science Red in Tooth and Claw*. Chicago: University of Chicago Press.
1982. *Darwinism Defended: A Guide to the Evolution Controversies*. Reading, MA: Benjamin/Cummings.
1986. *Taking Darwin Seriously: A Naturalistic Approach to Philosophy*. Oxford: Blackwell.
1988a. *Homosexuality: A Philosophical Inquiry*. Oxford: Blackwell.
1988b. *But Is It Science? The Philosophical Question in the Creation/Evolution Controversy*. Buffalo, NY: Prometheus.
1996. *Monad to Man: The Concept of Progress in Evolutionary Biology*. Cambridge, MA: Harvard University Press.
2003. *Darwin and Design: Does Evolution Have a Purpose?* Cambridge, MA: Harvard University Press.
2005. Darwin and mechanism: Metaphor in science. *Studies in History and Philosophy of Biology and Biomedical Sciences* 36: 285–302.
2017a. *Darwinism as Religion: What Literature Tells Us about Evolution*. Oxford: Oxford University Press.
2017b. *On Purpose*. Princeton, NJ: Princeton University Press.
2021a. The Arkansas creationism trial forty years on. *Karl Popper's Science and Philosophy*. Editors P. Zuzana, and D. Merritt, 257–76. Switzerland: Springer.
2021b. *A Philosopher Looks at Human Beings*. Cambridge: Cambridge University Press.

2022a. *Understanding Natural Selection*. Cambridge: Cambridge University Press.
2022b. *Why We Hate: Understanding the Roots of Human Conflict*. Oxford: Oxford University Press.
Ruse, M., and R. J. Richards, Editors. 2017. *The Cambridge Handbook of Evolutionary Ethics*. Cambridge: Cambridge University Press.
Ruse, M., and E. O. Wilson. 1986. Moral philosophy as applied science. *Philosophy* 61: 173–92.
Russell, B. 1945. *A History of Western Philosophy*. New York: Simon & Schuster.
1959. *My Philosophical Development*. London: Allen & Unwin.
Russell, B., and A. N. Whitehead. 1913. *Principia Mathematica*. Cambridge: Cambridge University Press.
Saini, A. 2017. *Inferior: How Science Got Women Wrong — and the New Research That's Rewriting the Story*. Boston: Beacon.
Salzberg, S. L., O. White, J. Peterson, and J. A. Eisen, 2001. Microbial genes in the human genome: lateral transfer or gene loss? *Science* 292: 1903–6.
Schelling, F. W. J. 1797 [1988]. *Ideas for a Philosophy of Nature*. Cambridge: Cambridge University Press.
Sebright, J. 1809. *The Art of Improving the Breeds of Domestic Animals in a Letter Addressed to the Right Hon. Sir Joseph Banks, K.B.* London: Privately Published.
Secord, J. A. 2000. *The Extraordinary Publication, Reception, and Secret Authorship of Vestiges of the Natural History of Creation*. Chicago: The University of Chicago Press.
Sedley, D. 2007. *Creationism and Its Critics in Antiquity*. Berkeley: University of California Press.
Seeman, T., L. F. Dubin, and M. Seeman. 2003. Religiosity/spirituality and health: a critical review of the evidence for biological pathways. *American Psychologist* 58: 53–63.
Selous, E. 1901–2. An observational diary of the habits – mostly domestic – of the great crested grebe (Podicipes cristatus). Continued as: An observational diary of the habits – mostly domestic – of the great crested grebe (Podicipes cristatus), and of the peewit (Vanellus vulgaris), with some general remarks. *Zoologist* 5: 161–83; 5: 339–50; 5: 454–62; 6: 133–44.
Sheppard, P. M. 1958. *Natural Selection and Heredity*. London: Hutchinson.
Sidgwick, H. 1876. The theory of evolution in its application to practice. *Mind* 1: 52–67.
Simpson, G. G. 1944. *Tempo and Mode in Evolution*. New York: Columbia University Press.
1949. *The Meaning of Evolution*. New Haven, CT: Yale University Press.
1953. *The Major Features of Evolution*. New York: Columbia University Press.
Smith, A. 1776. *The Wealth of Nations*. London: Strahan, W. and Cadell, T.
Smith, H. M., D. Chiszar, and R. R. Montanucci. 1997. Subspecies and classification. *Herpetological Review* 28: 13–16.
Smith, P. 2023. More congregations leave United Methodist church. *Tallahassee Democrat*, December 31, 8A.
Smocovitis, V. B. 1999. The 1959 Darwin centennial celebration in America. *Osiris* 14: 274–323.
Sonstroem, D. 1998. The breaks in "Silas Marner." *The Journal of English and Germanic Philology* 97 (4): 545–56.
Spencer, H. 1851. *Social Statics: Or, the Conditions Essential to Human Happiness Specified, and the First of Them Developed*. London: Chapman.

1852a. A theory of population, deduced from the general law of animal fertility. *Westminster Review* 1: 468–501.
1864. *Principles of Biology*. London: Williams and Norgate.
[1852b] 1868. The development hypothesis. Reprinted in *Essays: Scientific, Political and Speculative*. Editor H. Spencer, 377–83. London: Williams and Norgate.
[1857] 1868. Progress: its law and cause. *Westminster Review* LXVII: 244–67.
1876. Society an organism. *Popular Science Monthly* 9 (May): 1–11.
1879. *The Data of Ethics*. London: Williams and Norgate.
Sulloway, F. J. 1979. *Freud: Biologist of the Mind*. Cambridge, MA: Harvard University Press.
1982. Darwin and his finches: the evolution of a legend. *Journal of the History of Biology* 15: 1–53.
Tattersall, I. 2022. *Understanding Human Evolution*. Cambridge: Cambridge University Press.
Teilhard de Chardin, P. 1955. *The Phenomenon of Man*. London: Collins.
Templeton, A. R. 2013. Biological races in humans. *Studies in the History and Philosophy of Biological and Biomedical Sciences* 44: 262–71.
Tennyson, A. 1850. *In Memoriam*. London: Edward Moxon.
Theobald, D. L. 2010. A formal test of the theory of universal common ancestry. *Nature* 465: 219–22.
Thomson, W. (Lord Kelvin). 1869. Of geological dynamics. *Popular Lectures* 2: 73–131.
Tinbergen, N. 1968. On war and peace in animals and man. *Science* 160: 1411–18.
Tindal, M. 1730. *Christianity as Old as the Creation: Or the Gospel a Republication of the Religion of Nature*. Newburgh: David Denniston.
Toulmin, S. 1967. The evolutionary development of natural science. *American Scientist* 57, 456–71.
1972. *Human Understanding*. Oxford: Clarendon Press.
Tucker, G. M. 1994. *Birds in Europe: Their Conservation Status*. London: Birdlife International.
Tutt, J. W. 1890. Melanism and melanochroism in British lepidoptera. *The Entomologist's Record, and Journal of Variation* 1 (3): 49–56.
Van Wyhe, J. 2002. *Darwin Online*. Darwin-online.org.uk.
Veenendaal, M. V., R. C. Painter, S. R. De Rooij, P. M. Bossuyt, J. A. Van der Post, P. D. Gluckman, M. A. Hanson, and T. J. Roseboom. 2013. Transgenerational effects of prenatal exposure to the 1944–45 Dutch famine. *BJOG*. https://pubmed.ncbi.nlm.nih.gov/23346894/.
Von Bernhardi, F. 1912. *Germany and the Next War*. London: Edward Arnold.
Wallace, A. R. 1858. On the tendency of varieties to depart indefinitely from the original type. *Journal of the Proceedings of the Linnean Society, Zoology* 3: 53–62.
1889. *Darwinism: An Exposition of the Theory of Natural Selection with Some of Its Applications*. London: Macmillan.
Watson, J. D. 1968. *The Double Helix*. New York: Signet.
Watson, J. D., and F. H. C. Crick. 1953. Molecular structure of nucleic acids. *Nature* 171: 737.
Watts, I. 1707. *Hymns and Spiritual Songs*. London: John Lawrence.
Weiner, J. 1994. *The Beak of the Finch: A Story of Evolution in Our Time*. New York: Knopf.

Weismann, A. 1882. *Studies in the Theory of Descent*. Translator R. Meldola. London: Sampson Low, Marston, Searle, and Rivington.

1889. The duration of life. *Essays upon Heredity and Kindred Biological Problems*. Translators E. B. Poulton, S. Schonland, and A. E. Shipley, 1–66. Oxford: Clarendon Press.

Weldon, W. F. R. 1898. Presidential address to the zoological section of the British association. *Transactions of the British Association*, 887–902. Bristol.

1899. Natural selection and fortuitous variation. *Popular Science Monthly* 55: 141.

Wheeler, W. M. 1923. *Social Life among the Insects, Being a Series of Lectures Delivered at the Lowell Institute in Boston in March 1922*. New York: Harcourt, Brace.

1939. *Essays in Philosophical Biology*. Cambridge, MA: Harvard University Press.

Whewell, W. 1833. *Astronomy and General Physics (Bridgewater Treatise, 3)*. London: William Pickering.

1837. *The History of the Inductive Sciences*. London: Parker.

1840. *The Philosophy of the Inductive Sciences*. London: Parker.

1845. *Indications of the Creator*. London: Parker.

[1853] 2001. *The Plurality of Worlds. A Facsimile of the First Edition of 1853: Plus Previously Unpublished Material Excised by the Author Just Before the Book Went to Press; and Whewell's Dialogue Rebutting His Critics, Reprinted from the Second Edition*. Editor M. Ruse. London: Parker.

Whitcomb, J. C., and H. M. Morris. 1961. *The Genesis Flood: The Biblical Record and its Scientific Implications*. Philadelphia: Presbyterian and Reformed Publishing Company.

White, A. D. 1896. *History of the Warfare of Science with Theology in Christendom*. New York: Appleton.

Whitehead, A. N. 1926. *Science and the Modern World*. Cambridge: Cambridge University Press.

1929 [1978]. *Process and Reality: An Essay in Cosmology*. New York: Free Press.

Wielenberg, E. J. 2010. On the evolutionary debunking of morality. *Ethics* 120: 441–64.

Wiener, P. P. 1949. *Evolution and the Founders of Pragmatism*. Cambridge, MA: Harvard University Press.

Williams, G. C. 1966. *Adaptation and Natural Selection*. Princeton, NJ: Princeton University Press.

Wilson, D. S., and E. O. Wilson. 2007. Rethinking the theoretical foundation of sociobiology. *Quarterly Review of Biology* 82: 327–48.

Wilson, E. O. 1975. *Sociobiology: The New Synthesis*. Cambridge, MA: Harvard University Press.

1984. *Biophilia*. Cambridge, MA: Harvard University Press.

1992. *The Diversity of Life*. Cambridge, MA: Harvard University Press.

2012. *The Social Conquest of Earth*. Cambridge, MA: Harvard University Press.

Wilson, L. 1971. Sir Charles Lyell and the species question. *American Scientist* 59: 43–55.

1972. *Charles Lyell. The Years to 1841: The Revolution in Geology*. New Haven, CT: Yale University Press.

Wiltshire, D. 1978. *The Social and Political Thought of Herbert Spencer*. Oxford: Oxford University Press.

Winfield, N. 2023. It is a sin, as is any sexual act outside of marriage. *National Catholic Reporter*.

Wittgenstein, L. 1922. *Tractatus Logico-Philosophicus*. London: Routledge & Kegan Paul.

Woodger, J. H. 1929. *Biological Principles*. London: Routledge & Kegan Paul.

WHO (World Health Organization). 2006. *Sickle-Cell Anaemia*. New York: Fifty-Ninth World Health Assembly.

World Jewish Congress. 2015. Late Austrian scientist Konrad Lorenz stripped of doctorate for lying about Nazi past. www.worldjewishcongress.org/en/news/late-austrian-scientist-konrad-lorenz-stripped-of-doctorate-for-lying-about-nazi-past-12-5-2015.

Wright, G. F. 1910. The passing of evolution. *The Fundamentals: A Testimony to the Truth*. Two Christian Laymen. VII: I, Chicago: Testimony Publishing Company.

Wright, S. 1931. Evolution in Mendelian populations. *Genetics* 16: 97–159.

1932. The roles of mutation, inbreeding, crossbreeding and selection in evolution. *Proceedings of the Sixth International Congress of Genetics* 1: 356–66.

1945. Tempo and mode in evolution: a critical review. *Ecology* 26: 415–19.

Wynne-Edwards, V. C. 1962. *Animal Dispersion in Relation to Social Behaviour*. Edinburgh: Oliver and Boyd.

Index

Printed in the United States
by Baker & Taylor Publisher Services